Construction Documentation

• 2ND EDITION •

JOHN A. RICCHINI
JAMES J. O'BRIEN

Printed in the United States of America

Library of Congress Catalog Card Number: 91-606-58

ISBN: 1-55957-219-1

Authors

John A. Ricchini (1934-1987) served the construction industry for 30 years as a contract administrator for architects, contractors, subcontractors and construction managers. Mr. Ricchini worked as chief architect and office manager of Eagle 2000 Engineers and Design in Pierre, South Dakota. He was also a nationally renowned author and lecturer and taught at the University of Nebraska as an associate professor. His articles have been published in *The Construction Specifier, St. Louis University Law Journal, The Journal of Design and Construction, Metal Building Review, The Atlanta Format, On the Beam, The Nebraska Specifier* and *The Nebraska Blueprint*. He appeared before the Philadelphia Grand Jury and served on the panel of arbitrators for the American Arbitration Association. He was a member of The Construction Specifications Institute and appeared on its masthead because of his many contributions.

Mr. Ricchini received a B.S. in architecture from Drexel University in 1965 and an M.S. in construction management from New Jersey Institute of Technology in 1975. He was an independent consultant for the construction industry in the areas of contract administration, field administration, specifications, claims preparation, and safety administration.

James J. O'Brien has been in the construction industry, after Navy service in Korea, for 35 years as scheduler, estimator, value engineer, and construction manager. He has been chairman of the board of O'Brien-Kreitzberg & Associates, Inc., since 1972. Previously, he was president of MDC Systems and executive vice president of a similar firm.

Mr. O'Brien has had field experience as a project engineer for the Rohm and Haas Company at Philadelphia, and then with the Major Weapons Systems Division of RCA, followed by three years with Mauchly and Associates (founders of the critical path method).

He is the author of 10 books on topics involving the construction industry and has authored more than 80 articles and presented lectures to various organizations. He has appeared as an expert witness more than 20 times.

He has served as chairman of the board of Project Management Institute, is certified as a project management professional (PMP), and was elected to the PMI highest honor, the Fellow designation. In 1983 he received recognition from PMI for his contributions to project management.

He is a member of the Society of American Value Engineers and is designated as a certified value specialist (CVS). He serves as a director for the Lawrence Miles Foundation.

Mr. O'Brien is a registered professional engineer in New York, New Jersey, Pennsylvania, Georgia, and Connecticut. He is a Fellow of the American Society of Civil Engineers and was named "Distinguished Civil Engineer" by the South Jersey Branch in 1986. He served as president of that branch in 1986–87. In 1987–89 he served as New Jersey State section president.

He received a bachelor of civil engineering degree (master of engineering equivalent) from Cornell University.

Table of Contents

Preface—Second Edition

Working on this edition has been a bittersweet experience. Much of what John Ricchini put in the first edition is still current and viable information. In fact, my experience in the construction industry is that it tends to operate in cycles, and those things which are "givens" and "truths" at one point in time tend to fade, and then be rediscovered. Accordingly, it was my task as co-author of the second edition to identify and keep those portions of John's work which continue to be relevant and active today. A second task was to broaden the scope of the work.

James J. O'Brien

Introduction

The major purpose of this book is to inform the reader as to what to document, how to document it, and why. In the process of establishing these three major factors, we will spend most of our time discussing the "why's." As a matter of fact, several chapters will be dedicated to establishing some basic fundamentals and principles which are encountered daily in the construction process. Understanding of these basic fundamentals and principles will help establish the "why's" of documentation.

Someone may ask why the "why's" and not more emphasis on the "what's" and "how to's"? The primary reason for establishing and understanding the "why's" is to motivate the individual into proper and accurate documentation. Once a person knows why documentation is important and is therefore motivated to document, the "what's" and "how's" are relatively simple. However, on the other hand, if a person knows what to document and how to document it but does not understand the "why's," lax and even inaccurate daily efforts may result.

Why documentation? First, certain documenting activities may be required by the contract. Second, accurate and frequent documentation will minimize the cost of claim preparation should one ever have to go to court.

A case which exemplifies the high cost of claim prosecution is *E.C. Ernst, Inc. v. Manhattan Construction Co. of Texas,* 551 F.2d 1026 (5th Cir. 1977). This case required four years of preparation and 40 days of court hearings (spanning a period of four months). Forty-one witnesses were called to the stand, 1,056 exhibits were presented before the bench, and the case involved five litigants. In a preliminary hearing, the judge advised all five litigants to find another way to resolve this dispute knowing that there would be much time and cost involved in the preparation and that no one would win. The judge chided the litigants with the following remark: "I, who have had no training in engineering, had to determine whether or not the emergency generator system . . . met the specs when you experts could not agree. This is a strange bit of logic."

The construction industry is, of its nature, embroiled in many activities over long periods of time. Even when those involved conduct an effective documentation program, the possibility of arising disputes and filing charges is almost inevitable. With the writing of this book, we hope to convey the seriousness of accurate, consistent, and timely documentation to help prevent claims and, in the process, to motivate participants into a profit-making mode.

CHAPTER 1

Legal and Administrative Relationships

BASIS FOR CONSTRUCTION AGREEMENTS

§ 1.1 In the conventional method of project delivery, the owner signs a contract with an architect or engineer to design a building. The documents prepared by the architect/engineer become the contract documents which the owner will use in executing a contract with a builder. In the contract between the architect and the owner, the two parties negotiate all the terms and conditions of that contract. The two parties literally form that contract by communicating with each other. If one says, "I want four percent," and the other says, "I'll only give you 3 1/2 percent," then they must negotiate until they come to a meeting of the minds and settle the differences. If the owner says, "I want 40,000 square feet and I want it on this site," and the architect, after researching codes and zoning ordinances says, "I can't give you that many square feet on one floor on that site because of the side yard restrictions," then the owner has a decision to make. The owner must either go to two stories, or be unable to get the 40,000 square feet.

Once the two parties have come to an agreement, they sign the contract. When they sign the contract, they have bound themselves to each other under the terms and conditions agreed to in that contract. In the eyes of the court, these two people have now come into an adversarial relationship with each other, as a result of that contract. (See Figure 1-1.) An adversarial relationship is one in which the two parties are at "arm's length" with each other. In simple terms, this means that they are on opposing sides and do not trust each other. Can we really say that they do not trust each other? In terms of this written contract, in terms of each party carrying out the commitment made with the other we can say that they do not trust one another. If they trusted one another and were

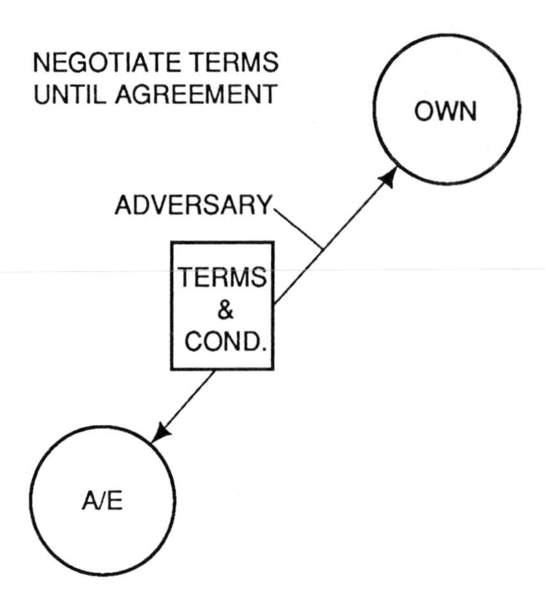

Figure 1-1

completely convinced that each would meet the commitment made to each other to the fullest extent, then there would be no need for a writing. However, in the real world there is an element of mistrust, or at least concern.

Some other basic principles of contract law are the factors that go into making a valid contract. There must be an offer and an acceptance between the two parties, and there must be consideration on the part of each party. This means that each party must give something to the other. If one party offers to paint the other party's house for $2,000, and the other party accepts the offer, then a contract is instituted. The offer was to paint the house and the acceptance was the owner's agreement to let it be done. The consideration on the part of the painter is giving paint and labor while the homeowner gives $2,000.

Another factor that affects the composition of a contract is that the two parties are left to themselves in negotiating and executing the contract. There is no arm of the law to referee the proceedings. If the two parties are of sound mind, have legal standing for themselves or the party they represent, and are not coerced into this agreement, then whatever they agree to becomes binding upon each other in the eyes of the law.

With the agreement with the architect settled, the owner proceeds to hire a contractor. Can the owner and contractor negotiate in a similar way? Can the two parties agree on the terms and conditions? They can negotiate the time and money but they cannot alter, via negotiation, anything that the architect/engineer has incorporated into the documents as part of the professional design.

What if the contractor says to the owner, "This architect that you've hired is looking for a design award by designing this building with a precast wall system, which is unique and very expensive. I can save you 10 percent, Mr. Owner, if you let me do it my way." Can the owner negotiate this contract requirement? No. The architect/engineer is a professional designer, recognized by and held accountable to society and the law for the integrity of the design.

Let's look at this contract from all three parties' positions: The architect, if aware of any such activity, must put a stop to it immediately. How does the architect do so? In writing. How serious must that writing be? If you were an architect, would you compose that letter or would you hire an attorney to compose it? In this situation, you should hire an attorney. That attorney will use the proper legal language to communicate to the owner (even though the contractor is initiating the change) because it is the owner who is violating the terms and conditions of the contract between the architect/engineer and the owner. Can the owner get a modification to the contract? Yes, if the owner goes through proper channels to secure the architect/engineer's approval. The courts have established a "standard of performance" for architect/engineers based on what is considered standard procedure by others in their professional community. One of the responsibilities of a professional designer is to see that no one changes the design. An attorney will use the proper legal language to preclude any and all claims, damages, etc., resulting from this action. Any and all claims doesn't mean just removing the architect/engineer from damages resulting from this particular change in the documents, but removing the architect/engineer from responsibility for the entire design. As you can see, this is very serious and needs the expertise of an attorney.

The owner runs the risk of losing any claim against the architect/engineer should any damages result from a poor design. The ramifications of the change may be far reaching. Should the foundation start to crack, the architect/engineer could defend by claiming that the cause of the cracking is the effect of the precast concrete wall panels on the foundation walls.

The contractor really runs the greatest risk of all. Should the building fail, or the owner suffer damages from the building, the contractor can be held liable for performing as an architect without a license.

Assuming that the owner and contractor come to an agreement regarding the length of time and amount of money involved in the contractor's offer, the owner then accepts the offer, and a contract is established. Please note this does not mean only when the contract is signed. When the owner accepts the offer of the contractor, the contract is established. It is important to understand that once the owner indicates acceptance either orally or by means of a letter of intent, the contract is in fact in existence. The signing of the contract confirms the oral agreement or letter of intent and is for purposes of record. (See Figure 1-2.)

The same contract principles apply to the contract between the owner and the contractor. When the contract is established, the two parties take on an adversarial relationship. They, too, are at arm's length with one another. When the contract between the owner and contractor is signed, another phenomenal thing happens. When the owner and the contractor enter into an adversarial relationship, as a result of their contract, the relationship between the owner and the architect becomes a fiduciary relationship. (See Figure 1-3.) The owner and the

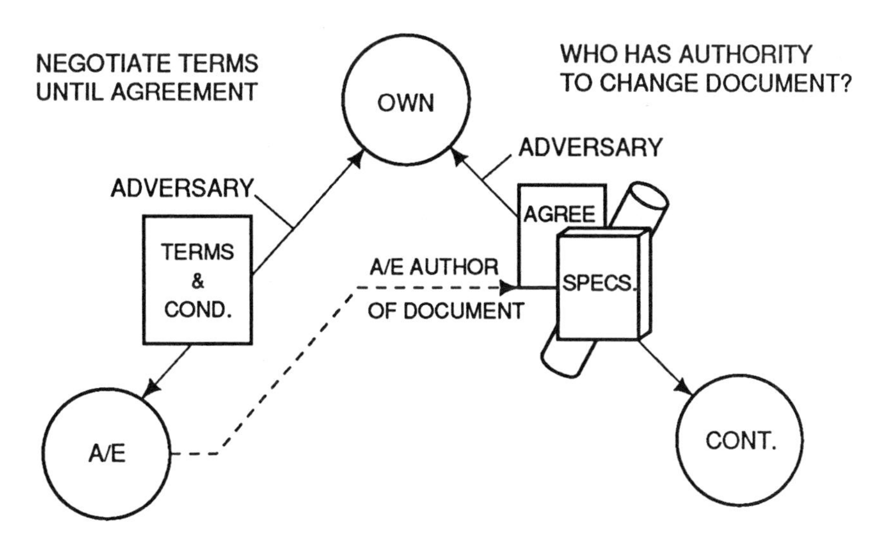

Figure 1-2

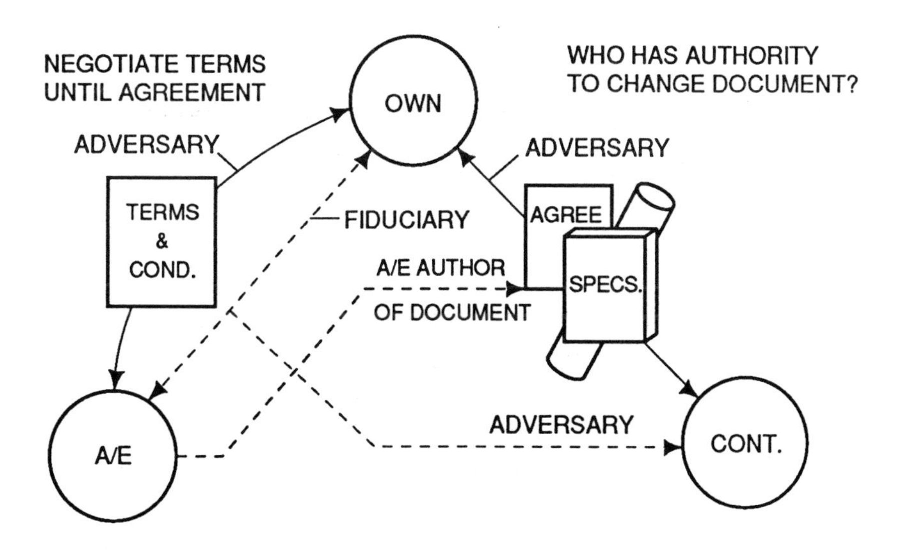

Figure 1-3

architect are adversaries as a result of the contract between them. The owner and the contractor are in an adversarial relationship as a result of the contract between them. However, once the contract between the owner and the contractor is signed, it puts the owner and the architect in a different relationship with each other and in their relationship with the contractor. Since the architect prepared the contract documents for the owner to use in securing a contractor, the architect and the owner are on the same side of the contract between the owner and the contractor. Take this one step further. The owner and the architect are now in a fiduciary relationship, one to another, in their adversarial relationship with the contractor. The architect and owner are still adversaries one to another as a result of their original contract. But they are fiduciaries to each other as a result of the contract signed between the owner and the contractor.

Is it important to go through all of that diagramming to determine who is in an adversarial relationship and who is in a fiduciary relationship? This is absolutely essential. Too many participants in the construction process do not realize how the law and the courts look at their respective roles. Too many times we relate to one another based on what we think are good psychologies, good personality traits, or good business practices. However, in the eyes of the court, it is what the contract binds the parties to within the framework of the law and the standards of the profession. It makes no difference how people think they should relate one to another to maintain a harmonious relationship with the other participants. What really matters is what the written contract, the courts, the written law, and the standards of your profession have dictated. Is this too strong for you? Am I promoting that everyone in the construction industry go about their business in an unusual way? Am I trying to discourage friendly relationships on a construction site? No. It's all right to be Mr. Personality. It's all right to maintain a harmonious relationship with all the other parties. However, you must do this within the framework of the above-mentioned controls.

Take a look at the condition that exists between the architect and the contractor. Is there a contract between the architect and the contractor? Is there a legally binding commitment between the architect and the contractor? The answer to the first question is a definite no. The answer to the second question is a very emphatic yes. Although there is no contract between the architect and the contractor, there are contractual duties that exist between these two parties, binding these two parties to each other as a result of the contract written by them with the owner. (See Figure 1-4.)

A simple example is the requisition for payment submitted by the contractor to the architect on a monthly basis. The architect owes the contractor a duty to evaluate that submission accurately and to process it within a reasonable period of time. The architect who fails in this duty could be found liable in the negligent performance of a contractual duty. Let's extend this example one step further. The architect should properly

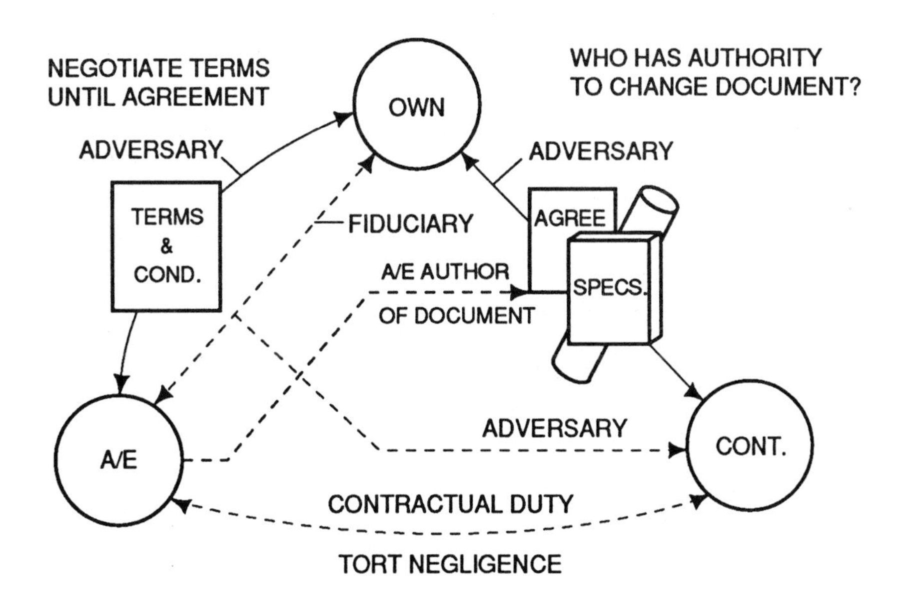

Figure 1-4

evaluate the request for payment and forward it on to the owner with the architect's approval and signature. Now, if for some reason, the owner does not process a check to the contractor for a long period of time, can the architect be held responsible for the owner's negligent performance? It is, sad to say, very possible that the architect could be liable for not monitoring the work of the owner in processing the payment back to the contractor. Can an architect be held liable for what appears to be negligence on the part of the owner? Remember that the architect is bound by the terms and conditions of the written contract. The architect can also be bound by the terms and conditions of the contract between the owner and the contractor as a third party to that contract. If the contract between the owner and the contractor states that the architect will monitor the work for compliance with the contract documents, then the architect is responsible for monitoring the work of both the contractor and the owner. If the owner is not performing and not processing a paycheck to the contractor is considered not performing, then the architect has the duty to advise the owner accordingly. The architect owes the contractor a duty regarding timely monthly payments made to the contractor, and this is binding on the architect.

A Florida Supreme Court case, *A.R. Moyer, Inc. v. Graham,* 285 So. 2d 397 (Fla. 1973), will solidify this issue. The court held that a state's general contractor who may foreseeably be injured or suffer economic loss caused by the negligent performance of a contractual duty by an architect has a cause of action against an architect. In simple everyday language, this means that the contractor has a right to go after an architect in a court for money lost as a result of the architect's negligent performance, even though there was no contract between the architect and the contractor. The finding of that case establishes the following areas of negligence on the part of the architect: the architect was negligent in the duty to prepare accurate plans and specs, in the preparation of corrected plans and specs, in the supervision of plans and specs, in the failure to award a certificate for payment upon completion of the project, and in the exercise of supervision and control over the contractor.

THIRD-PARTY RELATIONSHIPS

Construction's Eternal Triangle

§ 1.2 In establishing the legal relationship between the parties of a contract, especially in the construction industry's eternal triangle, it is beneficial to define the status of each party and the roles those parties play in the relationship.

Owner

§ 1.3 The owner is an individual, corporation, government agency or some other body in whom is vested the ownership, dominion, or title to property.

Contractor

§ 1.4 The contractor is one who undertakes to, or does, construct, alter, repair, add to, subtract from, improve, move, wreck, or demolish a building.

Architect

§ 1.5 The architect is one who forms or devises plans and designs and draws up specifications for buildings or structures and superintends their construction.

In a typical construction arrangement, a contract exists between the owner and the architect or engineer and one between the owner and the contractor. There is no written contract between the architect and the contractor. There is, however, a legal relationship, and this legal relationship stems from the contract written between the architect and the owner and the contract written between the owner and the contractor. In both these relationships, the basic function of the owner is to pay the architect for services rendered and pay the contractor for the building produced. The commitments of the architect to the owner and the contractor to the owner are somewhat more involved and necessitate further discussion.

The Owner/Contractor Relationship

§ 1.6 In the relationship with the builder, the owner agrees to pay the builder a mutually agreed upon amount of money in return for a building, in full compliance with the contract documents, adequate for its intended purpose. The owner also has the responsibility to provide the contractor with land and easements, engineering surveys, investigation reports of subsurface conditions, and copies of the insurance policies which include the interest of the owner, the contractor, and the subcontractors in the work.

The builder relates to the owner through the architect during the construction period. All information will pass to and from the owner and contractor through the architect. However, the contract documents shall not be construed to create any contractual relationship of any kind between the architect and the contractor. A.I.A. Document A201, General Conditions of the Contract for Construction, Article 1.1.2 (1987).

The Owner/Architect Relationship

§ 1.7 In the relationship with the architect, the owner agrees to provide compensation to the architect for furnishing professional services which include preparation of the contract documents, estimations of the costs, advice on the selection of contractors–and acting as the owner's representative in dealings with others associated with the project. The owner reserves the right to terminate the work should the architect fail to perform the contract in accordance with its terms through no fault of the owner. The owner has the responsibility to provide the architect with information regarding the site, the budget, and program criteria and to give prompt and thorough attention to all documents submitted by the architect. Also, when deemed necessary by the architect, the owner is to furnish and pay for the services of soils consultants, legal consultants, and auditing and insurance consultants.

The architect is responsible to the owner for the design of the building, including the structural, mechanical, and electrical systems as well as the architectural design. The architect is also responsible for assisting the owner in letting the building contracts, monitoring the work during the period of construction, issuing certificates for payment to the owner for approval of payment to the contractor for the work completed thus far, and issuing a final certificate declaring the building complete according to plans and specifications. In addition to all these responsibilities to the owner, the architect also functions as an arbitrator or quasi arbitrator in settling claims, disputes, and other matters in question between the contractor and the owner relating to the execution or progress of the work or the interpretation of the contract documents.

The Architect/Contractor Relationship

§ 1.8 Although no contractual relationship exists between the architect and contractor, there is a relationship which legally ties the two parties through their relationship with the owner. However, this relationship between the architect and the contractor through the owner did not take roots until well into the 20th century. Prior to this time the architect was protected against any liability to the contractor because of absence of privity.

Theory of Privity

§ 1.9 The evolution of the legal relationship between the architect and the contractor passed through several stages starting with the architect being held accountable for actions only through fraud or collusion. The architect in the role of quasi arbitrator between the contractor and the owner was immune from liability since the role as quasi arbitrator could affect the capacity to judge any dispute between the two parties. The contractor, was liable to those with whom there was a contract but not until the middle of the 19th century did the architect become liable to those parties with whom the architect had privity. However, as to a third party with whom the architect or the contractor did not have privity, there was no duty or obligation to that third party. Not until the middle of the 20th century did the architect and contractor become liable to a third party through a contractual relationship.

In order to better understand the stages through which the liability of both the architect and the contractor passed, we will present, in a chronological order, landmark cases and sample cases to help clarify the sequence and magnitude of each change. A landmark case is one in which the court clearly defines the issue, analyzes and sets aside extraneous matter, renders a decision based on a specific line of reasoning, and as a result, establishes a precedent for subsequent similar cases.

In the middle of the 19th century, the courts established that the architect and contractor were not responsible or liable to any party other than those with whom they had a contractual obligation. The state of privity between two parties established a legal contractual relationship and it was only through that relationship that an architect or contractor could be held liable. Several cases made strong statements to this effect.

In *Mayor of the City of Albany vs. Dunliff,* 2 N.Y. 165 (1849), the plaintiff was injured when a negligently constructed bridge collapsed. The court denied liability on the part of the builder because no connection between the wrong done and the person whom it is sought to charge for the consequences existed.

In *Curtain v. Somerset*, 21 A. 244 (Penn. 1891), the court upheld the no privity rule and stated that if one who erects a house or builds a bridge owes a duty to the whole world that the work contains no hidden defects, it would be difficult to measure the extent of the person's responsibility and no prudent person would engage in such occupations upon such conditions.

Many other cases were tried with lack of privity protecting the architect and contractor as well as any other person who was not in privity with someone suffering damages as a result of the negligence of the party against whom the action was brought. However, early in the 20th century, signs of erosion of the "no privity" rule appeared. This movement was third-party liability, which spread rapidly across the land.

Third-Party Theory

§ 1.10 The fall of "no privity" was long in coming and, even after serious inroads were made, cases heard as late as 1947 relied on and were successful in pleading "no privity." Third-party liability entered the scene in 1903 with a minor case and 13 years later was the theory in a landmark case in New York City, *Huset v. J.I. Case Threshing Machine Co.,* 120 F. 865 (8th Cir. 1903). In this case the court established that the seller was liable if the seller knew that the chattel was dangerous for its intended use, and that the use was a type inherently dangerous to human safety.

In *MacPherson v. Buick Motor Co.,* 217 N.Y. 382 (1916), Judge Benjamin N. Cardozo concluded that the duty did not arise out of contract, as many of the courts had held when the privity of contract rule had been applied. Rather, the duty arose when the manufacturer of the article knew that the article was to be used by persons other than the purchaser, and that the article was reasonably certain to put human life and limb in peril if negligently made. *See* Thornton and McNiece, "Torts and Workmen's Compensation," 32 *N.Y. Law Review* 1465 (1957).

The *MacPherson* case only extended liability to immediate purchasers. However, the courts did not waste much time before applying the *MacPherson* rule to members of the purchaser's family and even to casual bystanders. As a result of the *MacPherson* rule, the courts began considering the nature of the article rather than whether the parties were joined by contract in determining liability to remote users. If the court found that the manufactured product was inherently dangerous if negligently made, then the manufacturer could be found liable. However, the term "inherently dangerous" was difficult to define, and as a result, many cases were heard and decided either way, based on the courts' interpretations of the "inherently dangerous" finding.

Although the *MacPherson* rule brought about a whole new legal relationship between the remote user and the manufacturer, it did not affect the construction industry. In cases involving permanent real property structures in which the court found either the architect or the contractor negligent, unless the party damaged was in privity to either the builder or the architect, there was no liability. It was not until 1949 that the construction industry felt the effect of the *MacPherson* rule.

Third-Party Theory–The Construction Industry

§ 1.11 In 1882, in New York City, a court applied third-party liability in the construction industry for the first time. However, the case involved a moveable scaffold which might be considered a manufactured article and not an integral part of a building structure. It was not until the middle of the 20th century that the courts applied the *MacPherson* rule to permanent building structures. The following are sample cases where the *MacPherson* rule was applied, including the landmark case of *Inman v. Binghamton Housing Authority.*

In *Foley v. Pittsburg-Des Moines Co.,* 68 A.2d 51 (Pa. 1949), the court applied the *MacPherson* rule to those who build structures as well as to those who supply chattels. The Pennsylvania court stated

> [T]here is no reason to believe that the law governing liability should be, or is, in any way different where real structures are involved instead of chattels. . . . The principle inherent in the *MacPherson v. Buick Motor Co.* case and those that have followed it cannot be made to depend upon the merely technical distinction between a chattel and a structure built upon the land.

68 A.2d at 533.

In *National Security Corp. v. Malvaney,* 221 Miss. 190 (1954), The court extended the responsibility of the architect to include third-party liability to the contractor's surety.

The case of *Inman v. Binghamton Housing Authority,* 3 N.Y.2d 137 (1957), became a landmark in the application of third-party liability to the construction industry. This case involved a two-year-old infant who was injured when he fell off a porch in a government housing project unit. The infant's father brought the action against the owner of the housing project, the contractor, and the architect. The incident occurred some six years after the building had been designed. The lower court dismissed the complaint based on lack of privity between the plaintiff and defendants. The appellate court reversed and held that the plaintiff did have a cause of action against the third-party defendants. Counsel for the plaintiff argued that the *MacPherson* rule should be applied to structures on real estate as well as to manufactured articles of personal property because such structures are inherently dangerous to third persons if they are defectively designed. The court agreed that there was no visible reason for any distinction between the liability of one who supplies a chattel and one who erects a structure.

Another principle of law which emanated from this case is that once the building or structure is accepted by the owner and there is not present a latent defect or hidden danger, there is no liability to third persons. This portion of the decision created what is now referred to as the patent-latent test. Simply stated, this test determines whether the danger which caused the damage was latent (hidden) and therefore beyond the control of the building owner or patent (readily seen upon reasonable inspection), and therefore preventable by maintenance of the building. If the danger was latent, the responsibility fell upon the architect or contractor. If the danger was patent, then the responsibility fell upon the owner.

Needless to say, the *Inman* case with its patent-latent test stirred much controversy in the courts and it was not long before the *Inman* rule was under attack.

Others extolled the decision which extended the application of the *MacPherson* rule to the building of structures on real property, but little was said in favor of the decision which applied the patent-latent test of liability to dismiss the complaint against the architect, the builder, and the owner. Immediately after the *Inman* case, several other cases were decided which had some impact on the architect's legal position.

In *Day v. National U.S. Radiator Corp.*, 128 So. 2d 660 (La. 1959), the lower court held that the design professional was liable for the death of a worker as a result of negligent inspection. On appeal to the highest court of Louisiana, the decision was reversed stating the supervising architect had no duty to protect third parties who are injured as a result of equipment owned and controlled by the contractor.

In *Erhart v. Hummonds,* 334 S.W.2d 869 (Ark. 1960), three workmen were killed by the collapse of a defective wall of which the architect had knowledge. The architect's contract contained a provision which gave him the authority to stop the work whenever it was necessary to ensure the proper execution of the contract. The court held that the issue was not whether the architect had breached a duty to the owner but whether he had breached a duty to the workmen arising out of the safety provision of the contract.

As witnessed from the above cases, even though the *Day v. National U.S. Radiator Corp.* case finally was decided in favor of the design professional, the legal theory rendering the architect and contractor liable to those outside their contracts, had been established.

Strict Liability Theory–The Construction Industry

§ 1.12 The theory of strict liability holds a defendant liable without determining fault. Whether the defendant exercised care or negligence, good or bad faith, knowledge or ignorance, the courts look beyond the acts or behavior of the individual and look at the nature of the cause of the damage. In the landmark case of *Schipper v. Levitt & Sons, Inc.,* 207 A.2d 314 (N.J. 1965), the court recognized the fact that the principle of strict liability had already been effected in the manufacturing world, and just as the *MacPherson* rule found its way into the construction industry, so did the principle of strict liability. In the opinion handed down by the court is the following statement:

> We consider that there are no meaningful distinctions between Levitt's mass production and sale of homes and mass production and sale of automobiles and that the pertinent overriding policy considerations are the same. That being so, the warranty or strict liability principles . . . should be carried over

> into the realty field, at least in the aspect dealt with here.

207 A.2d at 325.

Although the *Schipper* case found a builder-developer liable under the principle of strict liability, three years later the court extended the rule to architects and engineers, regardless of the number of houses which may have been built and sold. Once again the construction industry was faced with an expansion of professional liability which extended not only to the contractor but to the design professional as well.

QUASI ARBITRAL ROLE

Immunity of Arbitrator

§ 1.13 The word "immune" is defined in *Black's Law Dictionary* as an exemption (as from serving in an office), a freedom from duty or penalty, or a particular privilege. Although the word immune is not used in the language of the A.I.A. General Conditions, the inference is clear in paragraph 4.2.12:

> Interpretations and decisions of the architect will be consistent with the intent of and reasonably inferable from the Contract Documents. . . . When making such interpretations and decisions, the Architect . . . will not show partiality to either, and will not be liable for results of interpretation or decisions so rendered in good faith.

As an arbitrator, the architect must act impartially when rendering a decision in disputes between the owner and the contractor, despite a conflict of interest inherent in the architect's relationship with the owner. Although the contract between the owner and the architect assumes an adversarial "arm's length" relationship, those same two people also have a fiduciary relationship to each other in their contractual relationship with the contractor. Nevertheless, the architect must render an unbiased decision, in keeping with contractual obligations, even if that decision means going against the owner. Obvious favoritism to the owner may prompt the contractor to sue the architect for collusion.

Regarding the architect's liability, the courts have granted the architect immunity when the architect acts as arbitrator between the owner and contractor. The reason for this grant of immunity was set forth in *Lundgren v. Freeman,* 307 F.2d 104 (9th Cir. 1962). The court said the architects are immune from suits, since there would otherwise be a real possibility that their decision will be governed more by fear of litigation than by their own unfettered judgment as to the merits. To avoid the risk of being held liable in such disputes, the design professional must clearly establish that he or she did not contribute in any way to the damages of the two parties at dispute.

In *Lundgren v. Freeman,* the architects were shielded from actions brought by the contractor because of their role as arbitrators. However, the court stated: "If [the contractor] could prove that the architects did anything to his (the contractor's) damage that was not within the scope of their authority as agents or quasi-arbitrators, he [the contractor] could recover such damages as naturally flow from the action." *Id.* at 119

Although the extent of immunity appears to be clearly defined in the contract language and by the courts, architects have attempted to plead "immunity as arbitrator" in areas where they have no such standing. The architect, in the role as arbitrator, is only immune from private actions against the architect for damages resulting from acts performed in this capacity. If the architect acts outside of that role, he or she cannot claim to be immune from liability.

AGENCY RELATIONSHIP

§ 1.14 In a typical agency relationship, there exists three parties: agent, principal, and third party (as opposed to two parties in a contract relationship). These parties must come together to form an agreement with each party fully aware of his or her role, the extent of that role, and the relationship to each other. Two parties cannot form an agency agreement without the third party's acknowledgement.

Agent to Owner

§ 1.15 Most of our discussion will be on the unique role of the agent to the owner. A portion of a legal definition of an agent is as follows: "having assumed towards the owner the duty to act solely in the owner's interest." As you can see from this portion of the legal definition, the agent must have the owner's interest at heart. It is this specific relationship, this fiduciary relationship, that makes a real difference in the contractual relationship of the owner with any other party in the construction process.

It could well be that many, many years ago this relationship existed between the architect and the owner, but because of the tremendous liability that the architect has assumed in relation to the contractor, and the inability to defend against such liability, the architect has phased out of the role of agent. As a matter of fact, it is made very clear in the written contract between the owner and the architect, and the contract between the owner and the contractor, that the architect is the owner's representative and not the owner's agent. In the universally accepted A.I.A. Document A201, General Conditions of the Contract, Article 4.2.1 states the architect is "to act on behalf of the owner only to the extent provided in the Contract Documents."

Another definition of the word "agent" describes some of the duties of that role as "A person authorized by another to act for him. One entrusted with another's business . . . one who modifies, accepts performance of, or terminates contractual obligations." This is a very serious role and contains some very serious duties which emanate out of the role of authority. If one is an agent, one literally is in the same position as the owner. I like to differentiate the role of the agent from the role of the representative in this manner: The representative speaks for the owner. The agent speaks as though he or she were the owner. When the architect speaks as a representative of the owner, especially in matters concerning time and/or money on a construction project, the third party contractor will not act on the architect's directive until it is confirmed in writing with the owner's signature. On the other hand, when the construction manager, as the agent to the owner, directs the contractor to perform in an area which affects the time and/or cost of the project, the contractor responds accordingly.

In an agency relationship, there is a specific commitment established in writing between three parties. In a typical construction project, the three parties involved in an agency relationship are the owner, as the principal; the construction manager, as the agent; and the contractor(s) as the third party. The writing will include the establishment of this relationship and the extent of the authority of the agent. For instance, the writing may indicate that the agent construction manager is authorized to expend as much as one million dollars in extras without the owner's prior approval and signature. This is why the contractor will respond to an agent construction/manager's directive even though it has an effect on the cost and/or time. This relationship, which is established in writing, must be known to all parties involved. The owner and the construction manager cannot agree to this relationship without making third-party contractors aware of it. As a result of this writing, the relationship established is known in legal parlance as an "express agency." (See Figure 1-5.)

Not all agency relationships are express. An agency relationship can be established by means of "implications" and acted on by one of the parties. This agency relationship is known as "implied agency" or "apparent agency." In this instance, where there is no expressed agency relationship established, the construction manager or the owner may perform, or fail to perform, in an area, which would imply to the third party contractor that the construction manager is an agent to the owner. If, as an example, the written contract required that all change orders be authorized by the owner prior to the contractor performing any work on that change and the construction manager directed the contractor to proceed by issuing a letter authorizing commencement of the work, this act may constitute an implied agency. In another situation, the owner and the construction manager in a private

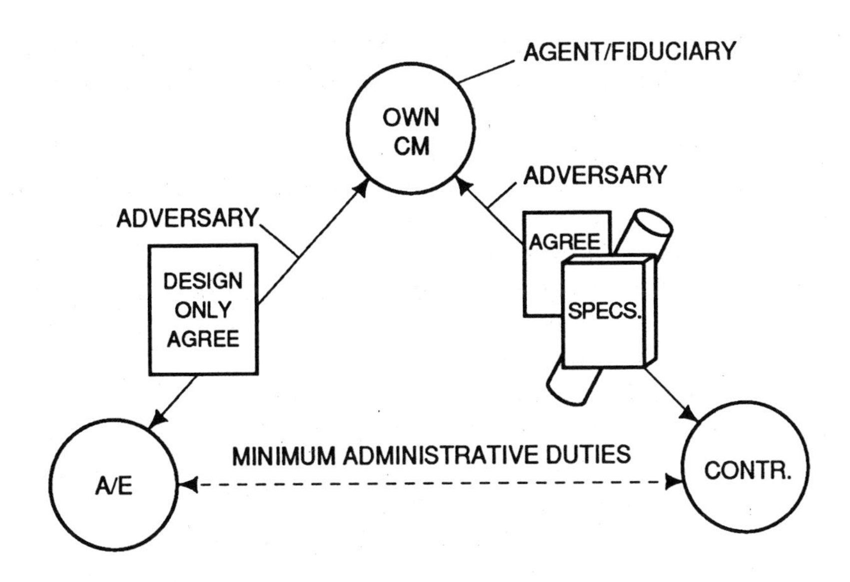

A/E-OWNER CONTRACT THROUGH CM AS AGENT TO OWNER
OWNER-CONTRACTOR CONTRACT THROUGH CM AS AGENT TO OWNER

Figure 1-5

discussion may agree to a change order, including the cost and extension of time, prior to discussing it with the contractor. When directing the contractor at a later time to proceed immediately with the work of the change order, the contractor questions the construction manager regarding the amount and extension of time. In the presence of the owner, the construction manager reemphasizes that the cost and time are not problems and to proceed immediately with the work. The contractor looks at the owner and finds no response either to the positive or negative. The contractor is not aware that these two have mutually agreed to proceed in this manner, and the contractor assumes that the construction manager has that kind of authority, and proceeds accordingly. When that construction manager acted, and the owner did not react, the contractor had no recourse but to proceed. When all this took place, the creation of an "implied agency" relationship took root.

In an outdated A.I.A. form, the instructions on the back of the form cautioned the architect regarding the possibility of becoming an agent. The instructions read as follows: "acts as owner's agent in dealing with others. . . . If not clearly defined by the writing . . . they may be determined in the courts by the laws of agency, customs of professional practice, acts of parties, circumstances of situation."

What these instructions are trying to establish is that the professional must be aware of practices, acts, and the circumstances of the situation which may establish an "implied agency" relationship. One factor is the definition of the word "implied" "where the intention is not manifested by an explicit and direct word, but is gathered by implication or deduced from the circumstances." Coupling the word "implied" with the word "agency," we have "one created by act of parties and deduced from proof of other facts."

What is significant regarding this role of agency and, in particular, the establishment of an implied agency relationship? Isn't the establishment of an agency relationship with the owner good for the industry? Isn't this what the owner has longed for to help resolve the problems experienced in the other methods of delivery? Certainly it is—if the parties mutually agree and are willing to assume the responsibilities that go with that role. However, there are many construction management arrangements whereby the construction manager does not want to be the agent. The construction manager wants to stay out of that role because of the tremendous liabilities that are contained in that role, and so it behooves the construction manager to be fully aware of this implied agency possibility. Wouldn't it be a disaster on the part of the construction manager if a written contract made it very clear that the construction manager was not an agent to the owner, and was therefore not responsible for the duties that come with that agency role, only to find out before the bench that the construction manager's performance in the field, or lack of performance in the field, established an implied agency relationship? As a result of the formation of that implied agency relationship, the construction manager is now liable as an agent, and the courts can hold the construction manager accountable for the damages done to the other party.

CHAPTER 2

Contract Documents

THE GENERAL CONDITIONS

§ 2.1 Typically the general conditions are published in printed form by professional organizations and trade associations such as the American Institute of Architects (A.I.A.), National Society of Professional Engineers (N.S.P.E.), Associated General Contractors (A.G.C.), and Associated Builders and Contractors (A.B.C.). Because these documents are prepared for every possible type and size of project across the country, the title "general" is truly applicable. Most projects, due to their unique concerns, will require that the general conditions be modified with supplemental or special conditions. This modification is accomplished by typing changes in a separate document and referencing them to the affected articles of the general conditions.

The general conditions have always been considered the legal portion of the "specifications." Justin Sweet refers to the general conditions in the preface to the second edition of his book, *Legal Aspects of Architecture, Engineering and the Construction Process* (2d ed. 1970), as follows:

> Courts treat such documents with considerable respect. These associations (A.I.A., N.S.P.E., etc.) are not governmental instrumentalities such as those traditionally classified as lawmakers such as legislatures, administrative agencies and courts. Yet realistically, such associations are "lawmakers" and their documents "law." The A.I.A. Document A511 "Guide to Supplementary Conditions" supports his statement and proposes that the "owner's legal counsel and insurance counsel review and approve all proposed provisions of the general and supplementary conditions."

Note that the A.I.A. does not recommend that the owner's attorney compose the conditions, but simply review and approve them.

When the architect incorporates the printed form into contract documents, the architect, in essence, becomes the author, a fact which has been established through the ages in the construction industry. Certain clauses such as the "indemnification," "liquidated damages," and "arbitration" clauses contain legal language beyond the comprehension of most architects. It is the owner's counsel who should dictate the terms of these clauses. However, the A.I.A. is aware of this situation and advises the architect in its "Guide to Supplementary Conditions" regarding such clauses.

There has been much talk regarding the "authorship" of the general conditions. The fact that they are printed and published by professional organizations certainly leads one to believe that the organizations are the authors. However, the courts see it differently. In the case of *Overland Constructors Inc. v. Millard School District,* 220 Neb. 220 (1985), the architect, as a third-party defendant, had to protect himself from liability to the school district, which claimed the architect neglected and refused to make adequate provisions in the project contract for the payment of charges to the utility district by the contractor.

The next fact to consider is the use of lawyers as the authors of the legal portion of the specifications. To have lawyers prepare the general conditions is, in theory, absolutely correct. The language of the general conditions is based on legal principles, policies, and methods with which only a lawyer is thoroughly familiar. However, from the practical point of view, the architect is the one who is most familiar with the procedures which must be followed in the contract administration of the construction project, and the architect is the one who is legally involved in maintaining compliance of such requirements.

To get a better perspective of who should be the author of the general conditions, one needs to ask some questions relating to the responsibilities placed upon the architect that emanate from the general conditions. Some of these questions are as follows:

1. Who administers the contract documents?
2. Is the owner's attorney ever mentioned in the general conditions? Does the attorney have a contractual duty to the parties of the contract?
3. Do courts hold the owner's attorney liable for the contract language? Is the owner's attorney considered the author of the general conditions, whether the attorney, in fact, did the composing?

Obviously the answers to all these questions will disclose to the reader that the lawyer, even when hired by the owner, as the composer of the general conditions, is not the one whom the courts look to as the author of the documents. In any dispute involving the contractor, the owner and/or the architect, the attorney for the owner will not be involved as the author of the general conditions.

The design professional needs the expertise of attorneys in the preparation of the general conditions of the contract. However, the architect is still responsible for its contents in the same way as the documents prepared by the structural, mechanical, and other consultants. This fact was also established by the courts when they held an architect liable for a collapsed wall erroneously designed by the architect's consulting structural engineer in the case of *Johnson v. Salem Title Co.*, 425 P.2d 519 (Or. 1967).

ADMINISTRATIVE REQUIREMENTS (DIVISION I)

§ 2.2 The Construction Specifications Institute (C.S.I.) has not only devised a system for having "a place for everything and everything in its place" when it created its 16-division breakdown, but it has taken care of another problem regarding confusion in the front-end documents. When C.S.I. created Division I, titled "General Requirements," what it really did was create a place for those concerns that are not legal or technical. More specifically, the topic headings in Division I are really administrative procedures and become the administrative link between the legal portion and technical portion of the project manual. In many instances, they also become the administrative link to the bid requirements.

A simple example of this administrative link theory can be demonstrated with the requirements for processing shop drawings. Obviously, from the legal portion of the project manual, it is required that no work be performed until the shop drawings for that particular work have been approved by the design professional. In the technical portion of the project manual, the individual sections will call out the specific technical requirements that are to be addressed in the shop drawings.

Going to the bid requirements, it is necessary that the bidder include the cost of processing shop drawings. Now we come to the administrative link and its role regarding shop drawings. We have seen how shop drawings must be addressed in the legal, technical, and bid portions of the project manual, but what is left to be addressed in the "administrative link" portion? Some very vital and costly concerns regarding shop drawings are the number of shop drawings required, the type of reproduction (sepia, blue-line, blueprint), and the extent of distribution (owner, design professional, and construction manager), all of which affect the cost of the project. If these administrative requirements are not spelled out accurately in Division I, then the contract would be lacking sufficient detail.

Another example of an administrative requirement that needs to be addressed in Division I is that of allowances. As we know, this subject is covered in the legal portion regarding the coverage of the allowance amount. Typically the amount represents the cost of the product, shipping, and taxes. The cost of installation is not included as part of the allowance sum. In the technical portion, the specific work that is covered by the allowance must still be addressed so that the contractor will know how much money to assign to the installation costs. In the administrative requirements, the areas affected by an allowance are itemized and the administrative procedures needed to monitor the allowance operation are established.

PRECEDENTIAL ORDER

§ 2.3 In the composition of the contract documents for a construction project, there is an order of precedence of one document over another. It is essential that the participant in the construction process understand this order and address daily activities accordingly. For instance, it would be foolish to approve a shop drawing against a contract drawing if there has been a change order issued which affects that contract drawing. In this instance, the shop drawing should be compared with the change order drawing and the original contract drawing in order to be accurate and complete in the evaluation.

The precedential order of construction contract documents is as follows, ranging from that which would be the most recent to that which would be the most remote. (See Figure 2-1.)

Change Order/Field Order

§ 2.4 The change order is a written document prepared by the design professional which affects the time and/or cost of the project and which must be signed by the owner, the architect, and the contractor before work relating to this change order is commenced.

A field order or minor change order can be an oral or written directive given in the field to change the original requirements without having an effect on the time or cost of the project. However, one must be cautioned about the possibility of a field order developing into a change order as a result of the requirements of the change. To avoid this situation, a document should be issued to all parties concerned, for their signatures, indicating that this is a field order and that there will be no effect on time and/or cost.

Contract Agreement

§ 2.5 This form will be signed by the parties to the contract and will establish the terms and conditions of the contract. In a construction project, since the contract is composed of many individual documents, this agreement form will list all of the documents that are included in the contract.

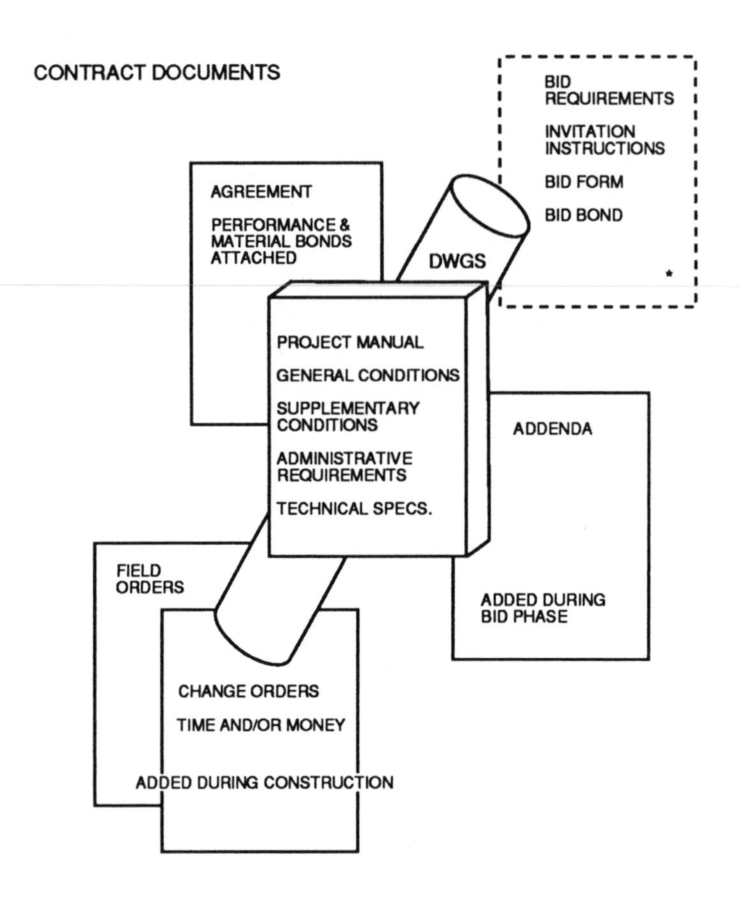

Figure 2-1

Addenda

§ 2.6 The addenda to the contract contains modifications to the original bid documents (including contract documents and bid requirements) during the period of time that the bid documents are on the market for solicitation of offers.

Supplemental (Special) Conditions

§ 2.7 Supplemental conditions are modifications to the general conditions, changing the contents of the general condition articles, or deleting from, or adding to the number of articles. These conditions are legal in nature and take precedence over the general conditions.

General Requirements (Division I)

§ 2.8 A listing of the administrative procedures required during the construction period of the project are contained in this section. Although the general requirements do not modify or alter the general conditions in any way, they sometimes clarify or supplement the general conditions and therefore take precedence over them.

General Conditions

§ 2.9 This section contains the legal requirements of the contract establishing the terms and conditions binding the two parties to the contract. These general conditions are "general" in nature and usually require modification or supplementing via the supplemental conditions or general requirements.

Specifications/Drawings

§ 2.10 These are the technical requirements for the material and methods of installation to be employed during the construction period. These technical documents are issued with the bid documents and become the construction documents once the contract is awarded.

Bid Requirements

§ 2.11 These are the bid procedures which every bidder must follow during the bid process. Many times, based on the writing of the instructions to bidders or the general conditions, these bid requirements drop into oblivion once the contract is signed.

INCORPORATION OF DOCUMENTS

§ 2.12 Documents outside of the contract documents may be incorporated by a reference clause. Typical of this procedure is the incorporation of the pre-printed form of the general conditions published by several professional organizations. Many design professionals, in issuing bid documents, include the general conditions by inserting a clause stating that by reference the specific set of general conditions are made a part of these contract documents. Although this is acceptable in the eyes of the court, it is a dangerous maneuver. Many times contractors, especially subcontractors, are not familiar with the contents of the standard general conditions and they place their bids without knowledge of the requirements stated in the general conditions.

Documents are also incorporated in the technical portions of the contract writing when a manufacturer's product is described and held as the standard for that particular piece of equipment or material. Any other product that would be substituted for the one specified must meet the written standards established by that manufacturer.

The law is also automatically incorporated into every contract. One cannot write a contract contrary to the law or in absence of the law. For instance, it is not necessary to state the contractor will meet all OSHA regulations, because OSHA is a law and is automatically incorporated as a requirement of the contract.

EFFECT OF STATUTES ON DOCUMENTS

§ 2.13 The laws or statutes are automatically incorporated in every contract writing. However, it is noteworthy to specifically point out several statutes which have a direct effect on the contract writing and its contents.

The statute of frauds is a statute which requires that contracts above a certain amount of money and extending over a certain period of time, must be in writing in order for the courts to recognize them as enforceable contracts. This does not mean that oral contracts are not legally binding, but that the courts can refuse to hear cases regarding oral contracts which violate the statute of frauds.

The Uniform Commercial Code typically affects products as opposed to services, and as a result, is not typically applied in the construction industry. However, since many products are installed in buildings, the code can be applied to those products.

What the code essentially says is that each product purchased is fit for its intended use or purpose. In essence, the code levies an implied warranty on all products used in the commercial market.

Workers' compensation laws are another concern for those in the construction industry. Although statutory law applies in every state in the union, there are conditions within some of the statutes that permit small organizations (those employing less than the number fixed by the state) to operate without workers' compensation insurance. When a subcontractor falls into this category, a statute typically requires that the prime contractor cover that subcontractor.

CHAPTER 3

Methods of Project Delivery

§ 3.1 For centuries the construction industry relied upon the conventional methods of project delivery, where the owner employs an architect to design a building and prepare the contract documents. Once the contract documents are ready, the owner then uses those documents to secure a building contractor to implement the architect's design.

Like other industries, the construction industry has its ups and downs. However, the construction industry has had some particularly severe "downs," and people in the industry began to wonder why there were so many problems. Owners of proposed buildings had the most to lose, and sophisticated owners started to demand that the industry render them better service, higher quality performance, less expensive methods, and shorter periods of time for completion. As a result of these pressures, both inherent and applied, the industry responded with innovative contracting methods in an effort to eliminate some of the problems and give the owners their due.

To compare these innovative methods, it is necessary to start with the conventional method, for in understanding the conventional method, greater appreciation can be derived for the more innovative methods.

CONVENTIONAL METHOD

§ 3.2 The conventional method of delivery can be divided into two major types: the first involves a single prime contract, and the second involves multiple prime contracts. The area of multiple prime contracts can be further refined into two segments: the first has to do with the standard process of employing multiple primes, and the second has to do with employing multiple primes but assigning or designating one of those primes as the coordinator.

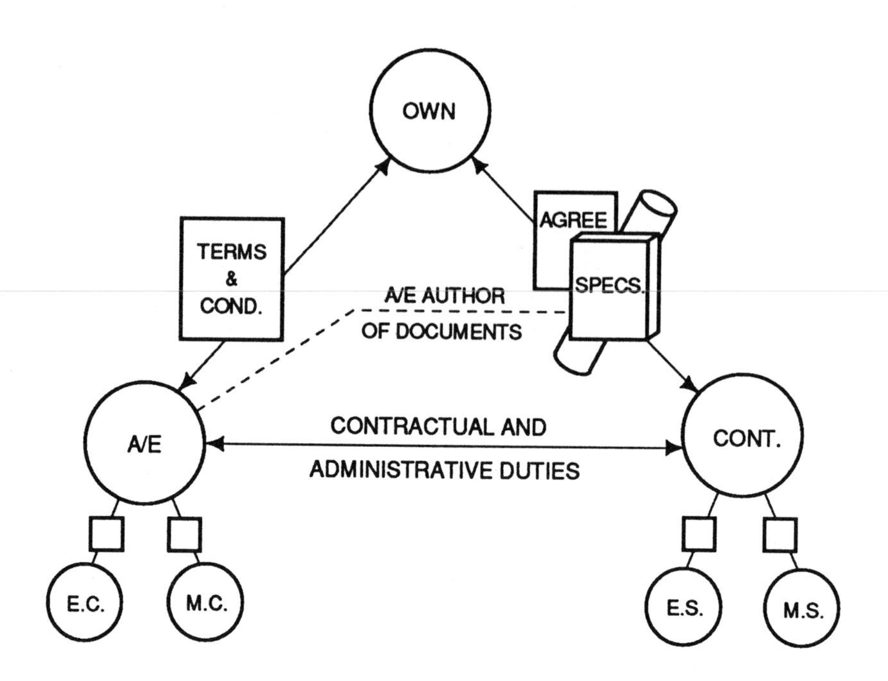

Figure 3-1

Single Prime Contract

§ 3.3 In a single prime contract situation, the owner selects one prime contractor (considered prime because the contract is directly with the owner), and the prime contractor in turn selects many subcontractors and suppliers to perform the work required for the project. In this arrangement, the relationship between the three major participants (the owner, the architect, and the contractor) places the architect as the representative of the owner, having authority over the contractor during the construction phase. The major role of the architect during this phase is to observe the activity of the contractor and the subs and make certain that the work is in compliance with the architect's design. (See Figure 3-1.)

One concern that the owner has, or should have, in this single prime contract arrangement is that the subcontractors may perform major portions of the work for which the owner has to pay a second layer of overhead and profit charged by the prime contractor. Another concern is that a conflict can arise in the relationship between the architect and the contractor. The architect, trained as a theoretician, can have difficulty communicating ideas to a very practical-minded contractor. The architect wants to maintain the aesthetics of the design, while the contractor wants to build a solid building to remain for years, complete the job in as short a time as possible, make as large a profit as possible, and move on to the next project. Working with these sometimes conflicting goals can make sparks begin to fly. Unfortunately, it's the owner who pays the price. These concerns in the single prime conventional method of delivery have caused owners to consider other approaches to contracting.

Standard Multiple Prime Contracts

§ 3.4 In the standard multiple prime conventional method of project delivery, an owner secures an architect to design the building and prepare the contract documents. However, the owner requires the architect to prepare the documents in such a fashion that they can be divided into several prime contract packages. The owner then selects contractors for each of these major portions of the construction project. Typically, these portions include the general contracting (commonly known as the architectural portion of the building), plumbing, heating, and electrical portions. On more sophisticated projects, the number of packages may be extended to include the elevator/escalator portion, site utilities, telecommunications, structures, and other major components of the project. Each of these primes, in turn, hires subcontractors and suppliers to perform specific tasks within their disciplines. (See Figure 3-2.)

The relationship between the three major participants (again, the owner, the architect, and the prime contractors) is one in which the architect, as representative of the owner, has authority over the contractors during the construction phase. The architect typically will monitor the work of each of the

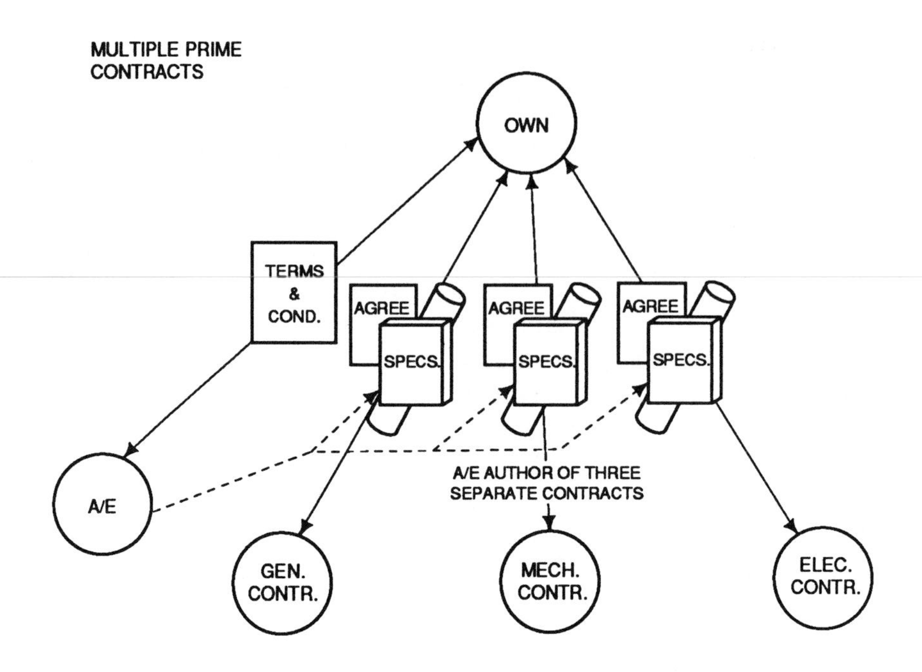

STANDARD MULTIPLE PRIME CONTRACTS

Figure 3-2

prime contractors and their subcontractors to guarantee that the work is in compliance with the contract documents. Due to the increase in the number of primes, and the possibility of overlapping areas of responsibility, the architect has a much more difficult job in monitoring the work and discerning who is at fault when problems arise.

The major problem in a multiple prime conventional method of delivery is the lack of coordination among the contractors. In the single prime arrangement, the single prime contractor has power to coordinate the work of all subcontractors and suppliers. If any of the subcontractors or suppliers does not cooperate, payment is withheld until order is resumed. In the multiple prime arrangement, each prime contractor has authority over its own subcontractors but does not have any authority over the other prime contractors. If their work overlaps, and there is conflict in either the time of performance or the placement of personnel, there is no one to settle the dispute. By standard contract language, the architect is not required to settle these disputes between primes, but is usually required to settle only disputes between the owner and the individual contractors. This is an area where the architect must be very careful not to exceed contractual authority. The architect resolving a dispute between prime contractors may assume the role of coordinator. Should that role evolve, the architect then takes on all the responsibilities of coordination.

Multiple Prime Contracts with Designated Coordinator

§ 3.5 Because of the serious problems regarding coordination that exist in a multiple prime situation, some owners have instituted a designated coordinator. This is accomplished by indicating in the contract and instruction to bidders that one of the prime contractors is designated as the coordinator, and all the other primes must adhere and comply with that requirement. In this arrangement, once again, the owner selects an architect who prepares the multiple prime packages for the owner. The owner, in turn, secures multiple primes, with the designated prime in the role of coordinator. (See Figure 3-3.) The architect is in the position of authority over the contractors during the construction phase and monitors their work for compliance with the documents. A major problem with the designated coordinator system is that although the designated prime contractor has the authority, via the contract, to coordinate the other primes, there is no means of enforcing directives.

DESIGN-BUILD METHOD

§ 3.6 Arthur Kornblut is an attorney from Washington, D.C., who has written many articles regarding the legal aspects of construction and was very instrumental in the formation of some of the A.I.A. publications. He defined the design-build method as a single legal entity holding direct responsibility (and liability) for project design and construction.

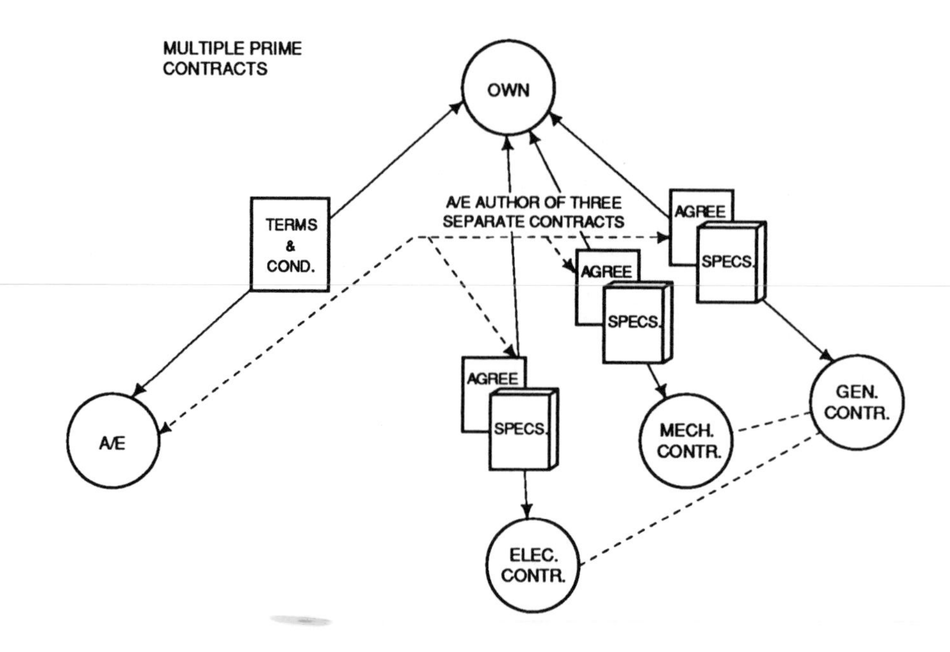

"ASSIGNED" MULTIPLE PRIME CONTRACTS

Figure 3-3

Design-build organizations can be formed in many ways. An owner may form an in-house design-build team. A contractor may hire a designer to complement the contractor's efforts. A designer may hire a contractor. A designer may form a joint venture. A contractor can employ a design professional, thereby having in-house design-build capabilities.

In the design-build method of delivery, the owner deals with one entity. There no longer need to be two entities (the architect to prepare documents and then the contractor to build from those documents). Rather, now there is one organization to accept the requirements and then meet those requirements.

On the surface, this arrangement appears to eliminate the conflict that exists between the architect and contractor.

However, in reality the conflict still exists. For the most part it is an undercurrent that simmers and stews on the back burner. The owner is not directly affected by it, but the strained relationship may have an indirect effect on the owner's pocketbook.

The design-build method is best employed on projects that are relatively simple and repetitive in nature. Warehouses, manufacturing plants, and commercial establishments all lend themselves to the design-build concept. Sophisticated building projects such as hospitals, arenas, and institutional-type buildings do not foster the design-build concept. The main reason is that the design-build entity usually employs the "fast-track" approach. In a fast-track situation, the designer prepares a small, up-front portion of the building package and then turns it over to the construction division to perform. While that portion of the work is under way, the designer continues with the next portion, and so on. The main reason for fast-tracking design-build projects is to compress the time frame, giving the owner a building far in advance of the time it would be completed under the conventional method of delivery. However, one can see that employing the fast-track system on a sophisticated building could be financially disastrous. The design professional must have a complete understanding of the sophisticated building before starting to put out packages for construction. Taking the time to do extensive research incorporating all of the complicated functions that a sophisticated building requires can nullify the time compression potential. (See Figure 3-4.)

DESIGN-CONSTRUCTION MANAGER METHOD

§ 3.7 This method of project delivery could be considered the public sector's answer to the design-build method. In the public sector, it is imperative that a government agency know the entire cost of a project prior to the awarding of a contract. Obviously, in the design-build method, where fast-tracking is employed, it is difficult to ascertain the cost up front. The design-build entity could give a "not to exceed maximum price," but the owner still would not know what the exact cost would be.

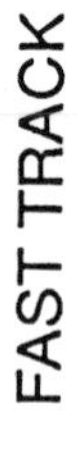

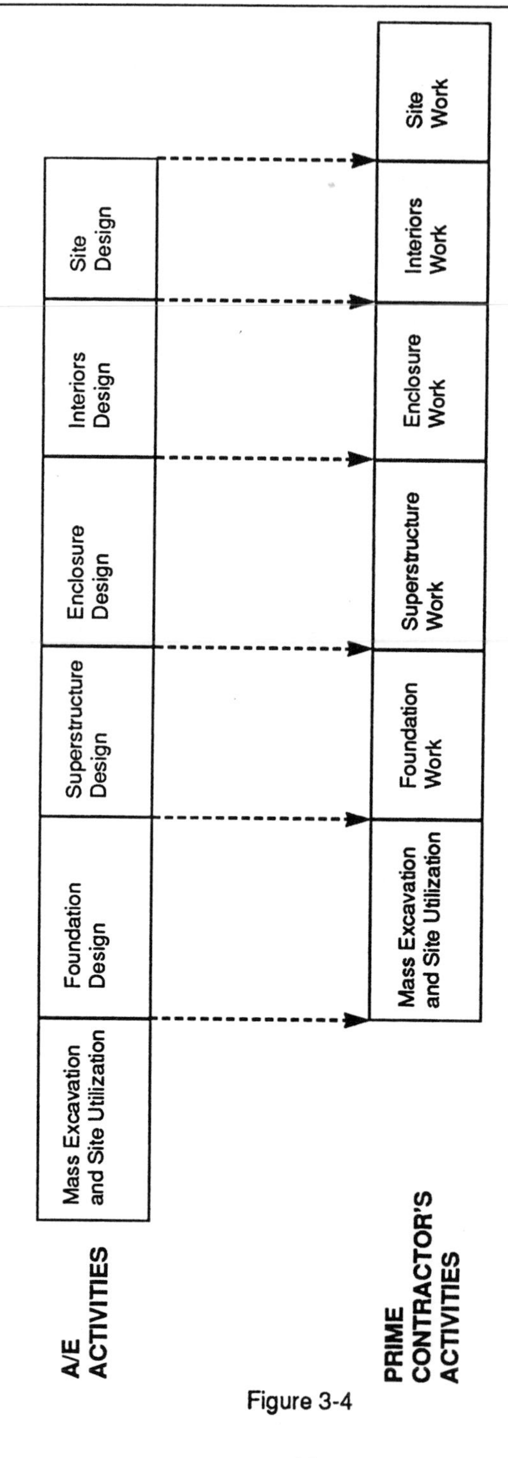

EACH ACTIVITY IS PERFORMED BY THE SAME PRIME CONTRACTOR. THE DESIGN PROFESSIONAL IS REQUIRED TO PROVIDE THE CONTRACTOR WITH THE DOCUMENTS FOR THE NEXT ACTIVITY BEFORE THE PREVIOUS ACTIVITY IS COMPLETE (ALLOW TIME FOR ESTIMATING AND PURCHASING). LONG LEAD ITEMS ARE CONSIDERED UP-FRONT TO MAINTAIN CONTINUITY OF SEQUENCE.

Figure 3-4

In the design-construction manager method of delivery, the public sector owner can benefit from both design and construction consultation during the design phase; in some instances this could be a major benefit to the building design. The construction manager can lend very valuable expertise to the design professional, leading to the prevention of some serious claims during the construction phase.

The design-construction manager method of delivery can also employ the counterpart of fast-tracking—a system appropriately labeled "phased construction." (See Figure 3-5.) Phased construction allows the designer to prepare contract documents for the early portions of the project (mass excavation, foundations, superstructure, etc.), which can then be put on the market for bidding and awarding of contracts while the designer continues with the subsequent portions of the building. This system will also compress the time frame for a project, and it will fulfill the public sector's requirement of competitive bidding.

Like fast-tracking, phased construction should only be applied to simple projects having a repetitive nature. More sophisticated buildings will require the conventional time for design and construction.

Note that fast-tracking is not used exclusively with the design-build method, nor is phased construction used exclusively with the design-construction manager method. Either one of these management tools can be used in other methods of project delivery, and certainly in the construction management method of delivery (discussed below). Moreover, phased construction is not limited to public work, nor is fast-tracking limited to private work.

CONSTRUCTION MANAGEMENT METHOD

§ 3.8 Construction management has many definitions emanating from various sources. These definitions vary because of the author's background, specific experiences with construction management projects, and the various applications, or misapplications, of the construction management

PHASED CONSTRUCTION

EACH PHASE IS AWARDED TO THE LOWEST QUALIFIED BIDDER. THE PROJECT MAY HAVE AS MANY PRIME CONTRACTORS AS IT HAS PHASES . . . EACH PHASE HAVING A DIFFERENT PRIME.

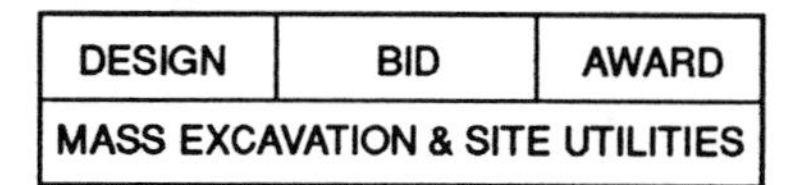

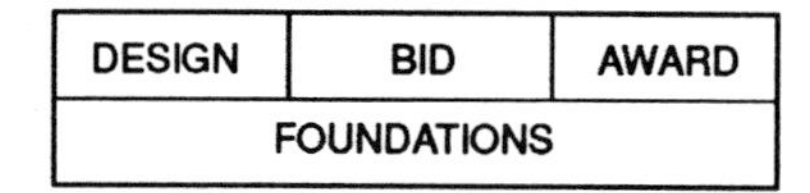

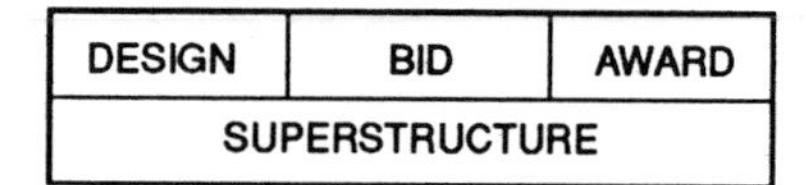

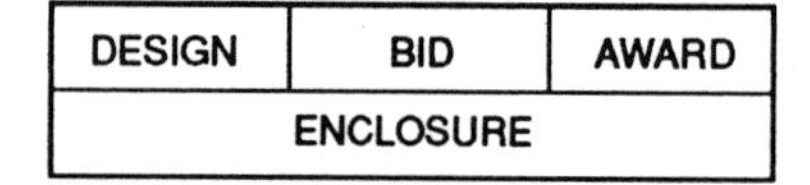

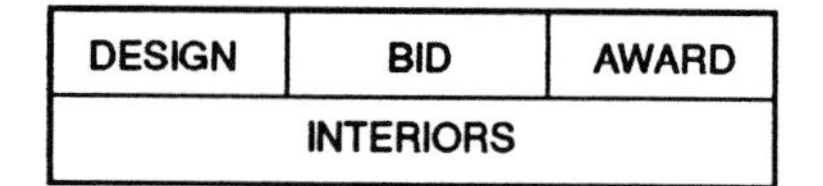

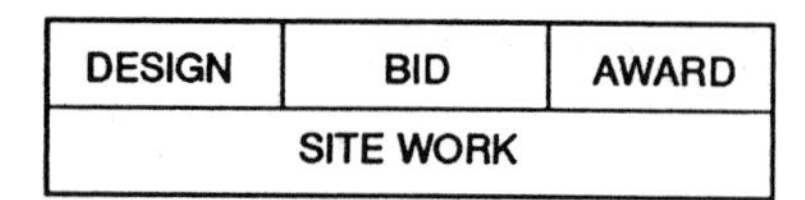

Figure 3-5

method of delivery. One unique method of construction management is called "professional construction management." Other methods of construction management tend to be variations of the traditional approaches. The unique factor in professional construction management is that the construction manager is the agent of the owner. Above all other requirements, considerations, arrangements, and legal relationships, it is the role of agent to the owner that makes the construction manager unique in the professional construction management form of construction management.

In the broad form of construction management, the construction manager can relate to the architect and contractors in several different ways: in some instances, as a co-administrator with the architect during the construction phase; in others, having authority over the architect and directing some of the architect's activities. In the more unique professional construction management form, the construction manager relates to the architect and contractors as the owner would relate to them. In the agency relationship, the construction manager literally takes the place of the owner and directs both the architect and contractors as the owner would direct them. It is in this arrangement that the owner hopes to avoid any conflicts present in the conventional, design-build, and/or design-construction management methods of delivery.

Some of the problems that exist in the construction management method of delivery, in the broad form, are the same as those experienced in all the other methods of delivery. In the professional construction management approach, however, the construction manager, through training, experience, and expertise, should be able to resolve problems prior to their developing to a level affecting the owner and/or the project.

TERMINOLOGY

§ 3.9 In the balance of this book, "construction manager" will be used to describe the party in charge of construction for the owner, which could be the project manager, construction manager, architect, or engineer.

CHAPTER 4

Bid Phase Documentation

§ 4.1 Are there concerns during the bid phase that require the participants to document their work and the activity of others? Absolutely. What is the bidding phase all about? Let's set the stage. This is a very dangerous area, depending on how the "contract documents" define the elements that make up those documents.

BID DOCUMENTS

§ 4.2 When the "bid documents" are matched with the "contract documents," do they both contain the same elements? The bid documents contain a portion called the bid requirements. It includes the invitation to bid, the instructions to bidders, the bid form, and the bid bond. They are the documents that literally tell the contractor how to conduct business during the bid phase and what forms are required. That's all they are designed to do. Now, if the contract writing does not include the bid requirements as part of the contract documents, then when the contract is awarded, those bid requirements disappear.

If A.I.A. Document A201, "General Conditions of the Contract for Construction," is used, the very first article defines the contract documents. In that definition, it states that the bid documents are not part of the contract documents unless they are specifically listed in the contract agreement form. The bid documents to which they are referring are the bid requirements (see Figure 4-1).

So if the contractor is using A.I.A. Document A201, and if he or she does not, on purpose, incorporate those bid requirements in the contract agreement form, or if the contractor does not supplement the General Conditions by modifying that statement, the contractor will lose the bid requirements at the signing of the contract. Is that dangerous? Yes. How dangerous? Very. When the bid requirements are excluded from the

BID DOCUMENTS	BID REQUIREMENTS	CONTRACT DOCUMENTS
Invitation	Invitation	
Instructions	Instructions	
Surety	Surety	
Proposal	Proposal	
Agreement		Agreement
General Conditions		General Conditions
Supplemental Conditions		Supplemental Conditions
Administrative Requirements		Administrative Requirements
Technical Specifications		Technical Specifications
Drawings		Drawings
Addenda		Addenda
		Change Orders

DISINTEGRATION

- Bid requirements disintegrate at signing of contract
- Transmit specific information pertinent to contract

PRECEDENCE

- Four corner rule
- Parol evidence
- Parol evidence rule

Figure 4-1

contract, the contractor can no longer require as part of the contract whatever was contained in the bid requirements. There may have been some unit prices or some critical dates included. The instructions to bidders state that the bid will be based on the evaluation of the documents and the site. If the instructions are dropped, the requirement for the bidder to go and evaluate the site has been removed. Serious? Very serious. There may have been some things on that site to be seen and included as part of the bid. What would happen if the contractor does not include a major element that could only be incorporated in the bid as a result of evaluating the site?

Consider how the courts may look at that situation. For instance, suppose there was a condition on the site that was part of the demolition requirement, at least in the mind of the designer, but was not called out on the demolition drawing or in the technical specifications. The condition is a foundation wall which is 14 feet deep, 400 feet long and 14 inches thick. Approximately three to six inches of the wall are exposed above the ground. The site is generally covered with weeds and garbage, so nobody ever saw the wall. The contractor begins mass excavation and runs into this wall. The contractor then makes a change order request. The construction manager rejects the request, stating that the contractor was to evaluate the site. The contractor is looking for $40,000 to remove the wall and so takes the claim to court.

The judge looks at the contract and reads it according to the "four corner rule." What does that term mean? The judge reads everything contained in the contract and nothing outside of it—for instance, newspaper articles that may have been written about the project. If the language of the contract is not favorable, the judge will interpret it and render a decision accordingly.

However, if there's an ambiguity and it's not clear what was meant, then the judge will look outside of the contract for clarification through parol evidence. What is parol evidence? Parol evidence is oral evidence or something written outside of the contract. In order to help clarify this ambiguity, the judge will permit, by the parol evidence rule, outside documents such as the instructions to bidders to come into the contract. Now, in the hypothetical situation described above, the bid requirements state that evaluation of the site is part of the basis for bidding. The contractor is saying this requirement is not part of the contract. The judge reads the first article in the contract. In the first article, it is very clear that unless specifically stated in the form of agreement, the bid requirements are not part of the contract. The judge agrees with the contract and states that the bid requirements are not part of the contract. It was not required that the contractor evaluate the site.

The lawyer representing the construction manager comes forward and says, "But, your Honor, you can use the parol evidence rule to bring in the bid requirements to help clarify the ambiguity." The judge says, "What ambiguity? There's nothing here that's ambiguous. There's no drawing illustrating the existence of that wall. There's no writing directing the contractor to remove the wall. There is no doubt regarding the removal of the bid requirements from the contract. Pay the man the $40,000 to remove the wall."

Now some folks will say that they won't let that happen on their projects. They will automatically include all of the bid requirements as part of the contract, or at least the instructions to bidders, and that will take care of that problem. In fact, many states have already done just that in their standard general conditions. The state of Kansas General Conditions state that the contract includes the bid documents or requirements as part of the contract documents. Does this method really eliminate the problem? In an actual account experienced on a HUD project, the bid form called for the project to be complete in 300 days. The instructions to bidders indicated 365 days. The instructions to bidders were expressly incorporated into the contract; the bid form wasn't. The owner had to give the contractor 365 days to complete the project. The owner really wanted the 300-day period and had included a liquidated damage clause to enforce this requirement. The inclusion of one of the two documents and not the other cheated the owner out of this requirement. Had both documents been included as part of the contract, there would have been an ambiguity.

The correct way to handle this matter is to do what the A.I.A. General Conditions suggest: to "specifically enumerate" the documents in the contract agreement form. To "specifically enumerate" means to read and evaluate each document, checking it for harmony with the other contract documents. They should not be included with the hope of avoiding one problem and in the process creating an even greater one.

PRE-BID CONFERENCE

§ 4.3 Is there an advantage to a pre-bid conference? First of all, if a pre-bid conference is needed, it should be a requirement of the bid process and included in the instructions to bidders. Otherwise, if the bidder doesn't show up for the pre-bid conference, the owner cannot utilize the results. When it is made a requirement, and even made a part of the pre-qualification statement, a bidder is disqualified for failure to attend. In public works projects, the pre-bid conference may not be able to be used as a means of pre-qualification. Government agencies will not permit the elimination of anybody from bidding based on such a requirement because everybody must have an opportunity to bid on a public works project.

The pre-bid conference should be an all-inclusive meeting. All of the bidders should meet at one time while the unique requirements for the project are discussed and demonstrated. Let's just say we're going to renovate an existing building, and we need to conduct a pre-bid conference because it's not easy to document the requirements in a writing or on a drawing. We are going to have to designate certain equipment for salvage, some to be removed and discarded, and some for repair.

The entire group should be taken through the building at one time. On one occasion the group was so large that the director of the institution conducting the meeting said, "Let's break into three groups. We'll have three leaders and tour the building at separate intervals." I was one of the bidders who walked through with one of the leaders who told us all about the building and what we had to watch out for.

The next day I went back on my own and visited with one of the other leaders (I used to be a member of the institution and I knew my way around). He took me through on a personal tour. He should never have done that, but then again the director should never have done what he did at the pre-bid conference. The leader told me of things that the first leader never even alluded to. One requirement was that four sections in a concrete foundation wall, approximately 14 inches thick, had to be cut to create huge openings. Remember, this is an existing building and the entire structure was resting on this wall. Do you think that's critical? The first leader never pointed this out. It was on the documents, but it was not clear and there was no indication that this was part of the main structure of the building.

Someone should take minutes of the meeting and the tour and document everything that is discussed. Copies of the minutes should be distributed to all those in attendance. They should be issued as an addendum and made a part of the documents. The pre-bid conference should not be just an incidental operation.

BID PREPARATION

§ 4.4 The bid period of a construction project should be treated as a separate entity, with great caution exercised on the part of all individuals concerned. Naturally, if the bid is not properly prepared, it will be considered nonresponsive and therefore rejected. Thus, all the work that goes into the preparation of a bid can be lost because of one negligent act on the part of the bid preparer.

A bid is nonresponsive when it does not meet all the requirements called for by the bid documents. A responsive bid is one that substantially conforms to all the requirements of the instructions to bidders. It must also substantially conform to all the technical requirements of the contract documents that are included as part of the bid documents. However, a bid that does not conform with the technical requirements is rendered nonresponsive only if the cause of the nonresponsiveness is more than enough to offset the lowest price. If the cause of the technical nonresponsiveness raises the lowest price past that of the next lowest bidder, then the bid is considered nonresponsive and therefore rejected. In legal parlance, this nonresponsiveness could be called a "material variance." This term simply means that the bidder has varied from the requirements and, as a result, the bid evaluator may declare the bid nonresponsive at the time of bid opening.

In addition to the bid being responsive, the bidder must also be responsible. A bidder who has the qualifications needed to perform the work of the project in the areas of technical ability, financial capability, and business integrity is considered responsible. These parameters which establish the responsible state of the bidder are not required until it is time to perform the work. In other words, the bidder does not have to possess these qualifications during the bid preparation period, but must have them at the time that the contract is ready for award.

The logic behind this thinking is to open the bid process to greater competition. Should these qualifications be necessary during the bid preparation period, there may be some bidders who will not be able to enter into the process. This requirement of responsibility flows down from the prime contractor to the subcontractors, and it is the prime contractor who is responsible for the subs' state of responsibleness. However, in some situations, the owner may require that the major subs establish their state of responsibleness directly to the owner.

With the advent of construction management, many construction projects are bid in the separate prime method of contracting. Under this method, the construction manager will solicit bids from many contractors, classifying them as prime contractors to the owner, even though many of these contractors functioned as subcontractors on previous projects. The logic behind this thinking is to minimize the overhead and profit that a prime contractor charges the owner for the work of the subcontractors. This situation also gives the construction manager direct control over the various prime contractors without having to go through an intermediary.

In the separate prime method of bidding, many owners will permit the primes to submit unified bids. This means that the prime can bid on one or more of the bid packages presented for solicitation. Naturally, this benefits both the owner and the prime contractor in that the cost of the separate prime contracts could be reduced if one prime were awarded several contracts. This cost reduction stems from the common use of equipment, labor forces, and mobilization.

However, there is a caution to bidders regarding unified bids. If the bid forms do not allow for entrance of this cost reduction and the bidder qualifies the bid, then the bid may be rejected. It is generally considered to be dangerous practice to qualify a bid, as precatory words are held to be unresponsive, which ultimately result in the bid being rejected. Another caution concerns the distribution of the cost reduction. Will the same reduction be present if the bidder is awarded three, four, or many more of the contract packages? Naturally, if unified bidding is permitted, then the preparer of the bid must be accurate in indicating percentage of cost reduction for the number of contracts awarded.

SUBCONTRACTORS' BIDS

§ 4.5 In more than a few instances, a subcontractor will seek to bind the prime contractor by asserting the formation of a contract. This attempt is usually made when the prime contractor uses the sub's quote in a bid and subsequently is awarded the prime contract. The sub in this instance feels that since the sub's quote was used and the award was granted, the sub should automatically have a contract. Obviously this is not so. The rules of contract pertain to the relationship between the prime contractor and the subcontractor the same as they do between the owner and the prime contractor. All the ingredients must be present before a contract is established. Those ingredients include an offer, an acceptance, consideration on the part of both parties, the capacity to contract, and a meeting of the minds.

If any one of these ingredients is absent, there is no establishment of a contract. The prime contractor's use of a sub's quote in the bid to the owner does not establish an acceptance of the sub's offer. Not until a written document establishing the agreement between the two parties is instituted can a contract exist.

Another claim made by subs regarding the institution of a contract is based on pre-award verbal negotiations. No matter how detailed these negotiations or how favorable the promise on the part of the prime, a contract does not exist until a written document is formulated by the two parties. Unfortunately for subcontractors, the reverse is not true. A subcontractor submitting a quote to a prime contractor can be held to that quote by the prime contractor should the prime contractor be awarded the contract. The logic here is that the prime contractor has relied upon the sub's quote in determining the figure bid to the owner. Once that figure has been accepted by the owner and a contract awarded to the prime, then the sub can be legally held to that quote.

The legal avenue used by a prime contractor to hold a sub to a quote is called promissory estoppel. Through promissory estoppel, the prime is able to stop the subcontractor from withdrawing a bid. Promissory estoppel is an equitable doctrine and can only result in an injunction directed at the subcontractor, and no damages can be assessed. If a prime makes a counteroffer to a sub's quote, then promissory estoppel cannot be considered.

In the Illinois case of *Gerson Electric Construction Co. v. Honeywell Inc.*, 453 N.E.2d 726 (Ill. 1984), the prime contractor (Gerson) relied on the subcontractor's (Honeywell) quote, verbally communicated via phone, in submitting a prime bid to the owner. In addition to the quote, the subcontractor had to submit a design scheme using its equipment as a requirement of the bid submission. After Gerson was awarded the contract, Honeywell refused to perform, insisting that it would have to more than double its original quote. Gerson withdrew its bid and the owner selected another contractor. Gerson then filed suit against Honeywell to recover $95,000 in profits it claimed it would have gained had it been the successful bidder.

The Illinois trial court ruled in favor of Honeywell, and the appellate court ruled that the contractor had no claim against

Honeywell for breach of contract. However, the appellate court concluded that Gerson might be entitled to its lost profits if it could prove that it had been injured because of the subcontractor's breach of the contract. To clarify the legal proceedings of the above case, one should note that Gerson did not apply promissory estoppel against Honeywell. Instead, it simply withdrew its bid from the owner and sued Honeywell for damages due to loss of anticipated profit. Had Gerson applied the legal principle of promissory estoppel, proving that it had relied on Honeywell's quote and had not made a counteroffer, then the courts may have placed an injunction on Honeywell from withdrawing its promise.

Some background on equitable doctrine will help to explain the court's decision in *Gerson*. Equitable doctrine emanates from a court of equity. The court of equity, contrary to a court of law, renders decisions based on what it considers to be just and equitable in spite of what the law may say.

The main theme in equity dictates that if a loss is to fall on one of two innocent parties, it should fall on the one who is in a better position to prevent the loss. Although there no longer exists the system of courts of equity, except in the state of Delaware, the theory and underlying principles of these doctrines have been adopted by many courts of law.

BID RIGGING

§ 4.6 The simplest way to explain bid rigging is that it occurs when two or more parties get together to outbid a third party. It can be accomplished without ever saying a word. As an example, one contractor may phone another contractor and say, "Don't bid on this job." The second contractor didn't even consider bidding on this project and simply ignores the request and goes about his other business. Later on, when the project is investigated and bid rigging is suspected, the second contractor may be implicated.

What should the construction manager do during the bidding process to abate bid rigging and to avoid being implicated? If, during the bid phase, there are any questions about the bid process, the construction manager should not answer them directly. The construction manager should not even indicate that the question is a valid question. If the construction manager permits the contractor to call with questions (and does not limit them to writing), the construction manager should simply say that if this is a valid question it will be covered by a written addendum.

When I was working as an architect on the Philadelphia Stadium, a manufacturer of a product called and said we had three different gauges indicated in the documents. The structural drawings showed one, the architectural drawings showed another, and the specifications showed a third. I was the administrator in charge of the bid phase. As I looked at the documents, I saw that he was right and I said so. What had I done? I had just given that particular individual an unfair advantage over the other bidders. At that time, no other bidder knew about that possible discrepancy. I may put out an addendum, but this manufacturer knows in advance of all the other bidders, and he can start putting his bid package together with this information incorporated. By the time the other bidders get the addendum, it may be the last day and they may not be as accurate in their hurry to put a bid together.

Let's look at another possibility. After I've taken the call I go to my structural engineer and he says there's no error. The gauges were meant to be different. One gauge is for the roof over habitable spaces, the other one is for the roof over the bleachers. What about the one in the specs? That has to do with the roof over the concession stands. Now what do I do? That bidder is now preparing a bid based on what I told him, and it was wrong. The construction manager should not indicate anything over the phone, and should not anticipate that an error, or supposed error, is legitimate until a written addendum is received.

Let me go one step further. I was called before the grand jury in Philadelphia for my role as spec writer, bid administrator, and resident architect on the stadium project. One of the issues in question was the gauge of the roof deck. Five people were indicted because of "alleged" collusion regarding that deck material. The five indicted included the manufacturer, contractor, erector, structural engineer, and a city administrator. Who do you think could have been a sixth member of the "team"? You're right! Yours truly could have been implicated because of my response to that phone call. The manufacturer indicted was the one who called that day.

Bid rigging is a violation of the Sherman Antitrust Act; that means it is a federal offense. Almost every case regarding bid rigging also involves mail fraud. Bid rigging is also considered a conspiracy, which is a criminal violation, a felony, punishable by fine or imprisonment, for which executives of very large firms in this country have gone to jail. Another fact regarding bid rigging is that the court will assess treble damages. That means that the damages suffered by the innocent party will be assessed by the court and then levied at three times the amount of the original assessment.

UNIFIED BIDDING

§ 4.7 In a typical construction management method of delivery, the bid documents are divided into many contracts whereby the bidders can bid on one, two, or as many portions as are let in the bid process. This arrangement of bidding one or more portions of the bid is covered in the instructions to bidders and is called unified bidding. Why would a construction manager do that? Why would a contractor want to bid all of the packages rather than one? Is there a benefit to the owner? Naturally, the owner is looking at the cost. Whichever "package" gives a low figure is probably the one to go with. Would it benefit a contractor? Naturally, the more portions a contractor has, the less coordination has to be provided.

An example of a coordination problem which could have been eliminated by a successful unified bid arose in *Peter Kiewit Sons Co. v. Iowa Southern Utilities Co.*, 355 F. Supp. 376 (S.D. Iowa 1973), a landmark case. The bids were put out under a unified bid procedure. There were six primes: two were for procurement—the boiler and the turbine—the other four were for construction—structural steel, general construction, mechanical, and electrical. Peter Kiewit bid on the general construction and was awarded the contract. Kiewit did not, however, have the structural steel portion, and lack of activity on the part of the structural steel contractor ultimately delayed Kiewit.

The steel contractor arrived late, had steel returned because it was improperly fabricated, and had an accident which damaged the work. The contract had a "no damages for delay" clause in it. This means that the owner will allow the extra time but will not pay any extra money. Kiewit was on the job several months after it was due to be complete. Kiewit sued Iowa Southern for extra money for the extended period of time that workers were employed on the project. Kiewit claimed the owner, via the project manager (construction manager), in this case Black & Veatch, did not coordinate the activities of all the other primes, namely the structural steel contractor who was causing the time extension problems for Kiewit.

The court examined the "no damage for delay" clause in the contract to determine its validity. The clause was very specific. It was clear language and was not ambiguous. It stated that the owner would grant the contractor a time extension but not any damages. Even though the conflicts and delays caused by the structural contractor were paramount to Kiewit's dilemma, the court held the "no damage" clause would be regarded as valid and enforced according to its terms. Kiewit lost the claim of $595,000.

REJECTION OF BIDS

§ 4.8 Most contracts, especially government contracts, contain the words "the owner reserves the right to reject any and all bids." When we talk about the right of the owner to reject bids, we're not simply talking about bids that have material variances which render them nonresponsive and therefore rejected. We are talking about the owner's right to reject any bid, even if the bidder is responsible and the bid substantially conforms to the bid requirements. The first question that comes to mind is why an owner should be granted such rights when it appears to be so unjust, especially if there is no cause for the rejection.

Volumes could be filled with the different jurisdictional requirements regarding bid rejection. In many jurisdictions, cause must be established, whereas in others, unless rejection can be proven arbitrary and capricious, no cause is necessary. In other jurisdictions, the rejection of a bid must be predicated on good faith and must be exercised promptly. Government authorities and owners who employ this right need to acknowledge the risk of violating the rights of the one rejected. In both the states of Texas and Connecticut, owners were held liable for "malicious interference with business expectance." In the case of *William F. Wilke, Inc. v. Department of Army of U.S.*, 485 F.2d 180 (4th Cir. 1973), the court permitted the bidder (Wilke) to sue and recover bid preparation costs because the owner had rejected the bidder's late bid. The court said that the bidder did not gain any competitive advantage by reason of tardiness. In the case of *Peru Associates, Inc. v. State,* 334 N.Y.S.2d 772 (1971), the contractor damaged by improper rejection of bids was entitled to recover reasonable profits, startup costs, and post-bid costs, but could not recover for bid preparation or pre-bid costs.

In protesting an award of bids, one must base the protest on the contract formation and not on the bidder performance. In essence, one must question the responsiveness of the bid as opposed to the responsibleness of the bidder. In the government arena, it is essential that a protester take immediate steps to file for an injunction against the governmental authorities awarding the contract. Often this action is required prior to the award of contracts, even to the point of seeking an injunction immediately after bid opening. The important factor is to register a complaint immediately. There will be ample time to justify and explain the basis of the objection. The idea is to prevent the agency from awarding the contract because it is the bid procedure and the alleged improprieties thereof that are in question. Once the contract is awarded, it may be too late to question the bid procedure.

BID WITHDRAWAL

§ 4.9 This could be a very dangerous area. The typical instruction to bidders language states that a bidder may withdraw before the bid opening date. The language does not say anything about whether a bidder may withdraw after the bids are open. Typically, when two people come together to form a contract, it is not until they agree on the terms and conditions and express that agreement that the contract comes into existence. One makes an offer, the other makes an acceptance, there is an agreement, and a contract exists. One of the other major factors regarding a legal contract is that no one can coerce another into a contract. If it is proven that someone coerced another into a contract, that contract will be rendered null and void. So in the construction industry, with its bidding system where bidders are solicited, bidders can withdraw before the bid opening. After the bid opening, they cannot.

What does the prime contractor do when a sub doesn't want to perform? Is that possible? It happens frequently. The sub says, "I'm sorry, but I can't do it. I've committed to other work while I was waiting." The contractor is now hurting because the bid was based on the bids from the subs, and now the subs are deserting the contractor. Can the contractor stop them?

The contractor can go to court and ask to apply the equitable doctrine of promissory estoppel. An equitable doctrine is one which emanates from the old court of equity and which generally judges matters on the basis of what is fair and just, in spite of the law. When promissory estoppel is applied, it permits the contractor to stop an act. It doesn't mean damages are assessed. It means the contractor can enjoin or stop the subcontractor from withdrawing. However, not all the courts allow equitable doctrines. Whether an equitable doctrine will be applied depends on the jurisdiction.

We have just talked about the relationship between a prime contractor and a sub. Can the same legal avenue be applied between the owner and a contractor? The best means to determine the legal ties between an owner and a contractor or bidder is through the courts. What are some of the courts' decisions?

Here is what happened in a case in Illinois. The bids were submitted to the owner. The owner accepted the low bid. The low bidder looked at the next lowest bid and found a tremendous amount of money "left on the table." The bidder came to the owner and asked to go over the bid, item for item, and see if something had been left out. The owner agreed. They did and there was nothing wrong. Then the prime bidder went to the subs and asked if they had missed anything. This bid was for an underground utility and the trenching sub noted that not only did they make an error, but it was so substantial that if they had to perform the job, they would go out of business. The prime and sub agreed not to go through with the contract. The prime went to the owner and asked permission to withdraw and not forfeit the bid bond. The owner denied the contractor's request and ordered him to perform.

The contractor took the owner to court. The court looked at the conditions presented to see if the error was a "material" error or one of "judgment." The court found that it was material error for the sub, after looking at the documents and finding some discrepancies, to never call for a clarification. Rather, the sub just based his bid on the discrepancies. (An error in judgment would have been if the owner of that subcontracting company said, "We need this job. Even though it's worth a million, let's take it for $800,000.")

The next condition the court looked at was the standard of care by the erring party. The party was able to produce documentation used in preparing this bid that showed it was a reliable and established firm that knew how to prepare bids. It was not a fly-by-night outfit prone to error. It had used an acceptable standard of care in the preparation of the bid.

The third concern of the court was whether it was unconscionable (against good conscience) to require performance. Again the evidence showed that this sub would have gone under if coerced into performing. The court said the coercion was against good conscience.

The final factor that the court had to determine was whether the owner would be injured by the withdrawal. If the owner has to go from the lowest bidder to the next lowest bidder, the owner is certainly going to lose money. Even if the owner was able to hold the lowest bidder's bond back, the amount wasn't as much as the difference between the two bids. However, this was a public bidding, and everything was published, including the budget. When the court looked at the budget, it said, "If you were to permit this bidder to withdraw, and were to go to the next low bidder, you still would not exceed your budget; therefore, you are not injured by the bid withdrawal."

A fifth factor, which was not in the Illinois case but exists in others, is that there must be prompt action in giving notice of the error. The bidder must not wait until the contract has been awarded and three months down the road say, "Oh, we made an error in our bidding." Notice must be done quickly. Withdrawal must be timely—the sooner the better.

It's important to know whether the prime contractor should be permitted to withdraw or not. Should the construction manager advise the owner to coerce the contractor to perform, or should the construction manager recommend that the owner release the bid?

BID EVALUATION

§ 4.10 Many factors must be considered in the evaluation of the bids. Is the bid bond included? How about the unit prices? What about the allowances, alternates, and options? Most people in the construction industry know what these last three topics represent; therefore, there is a tendency to gloss over them. However, there are some critical points.

An *allowance* is a sum of money set aside by the owner to remove a portion of the work from the competitive market. The hardware schedule is a good example. This must be watched in the bid phase, and in particular the construction manager has to make sure the architect properly incorporates this requirement into the bid documents. If the owner says to use $50,000 as an allowance for the hardware, does that mean the architect does not have to have a technical section in the specifications?

Some architects think that a hardware section is not needed because there is an allowance for it. This behavior indicates they don't understand what the word allowance really means.

As an example, consider a university where administrators want to maintain the same key locking system for a new building to match an existing building. To accomplish this, they may take that hardware portion of the work out of the competitive bid market so they can control whose product will be installed. That's about the only legal way it can be done in a public works operation. However, that $50,000 represents only the portion of the hardware that the administrators don't want to be competitive. The installation cost is still competitive. The university doesn't care who installs it, it just wants to make sure that what is installed is in harmony with the rest of the existing building system. Installation and handling costs will still need to be determined under the competitive bid process.

The construction manager must make sure that the architect clearly defines what the hardware allowance represents—that it represents the cost of the material, taxes, and shipping. The balance of the cost of the hardware portion of the work is still competitively bid. Since the balance is still competitively bid, then the architect must have a hardware schedule to show how many of these pieces are going to be put in and where they are going to be put. A technical spec section is still needed to define how the pieces will be installed. It must be complete for the bidder to put a price in for the competitive portion.

An *alternate* is a material or method selected by the owner, based on its cost, as an alternate to that material or method specified as a requirement under the base bid. For example, an owner has $2 million to do a building, but is not certain the amount will cover the cost of all the amenities. Preference is to cover the walls of all the corridors with tile, but the owner will accept vinyl wall covering if the cost is prohibitive. To resolve conflict, the architect/engineer will solicit alternate bids. The base requirement will be ceramic tile. The alternate will be wall covering. Should it be an "add" or "deduct" alternate? It's a deduct in this case.

An *option* is a material or method selected by the contractor from several choices presented in the documents; money is not a factor. There are times when options are presented to contractors because of other factors that are more important than money (minority programs, for instance). If it's more important that the contractor maintain a minority program, then the construction manager may give the contractor several options as to how to perform the work so the contractor can meet the minority program requirements.

APPENDIX 4A

RECORD OF PRECONSTRUCTION CONFERENCE

PAGE 1 OF 4

NAME OF OWNER	ADDRESS (Including Zip Code and Telephone Number)
NAME OF ARCHITECT/ENGINEER	ADDRESS (Including Zip Code and Telephone Number)
NAME OF CONTRACTOR (FIRM)	ADDRESS (Including Zip Code and Telephone Number)
LOCATION OF CONFERENCE	DATE

SUBJECTS TO BE DISCUSSED

1. Identification of Official Representatives of Owner, Engineer, Contractors and Other Interested Agency:

OWNER ______________________ ENGINEER ______________________

ADDRESS ______________________ ADDRESS ______________________

CONTRACTOR ______________________ OTHER ______________________

ADDRESS ______________________ ADDRESS ______________________

2. Responsibilities of Engineers and Architect
   (Does not "supervise" the Contractor's employees, equipment or operations):

3. Responsibilities of Owner's Governing Body (Actual Contracting Organization):

4. Responsibilities of Contractor (Review Contract Terms):

5. Responsibilities of Any Other Agency Contributing to the Project:

6. General Discussion of Contract:
   A. Alternative Specifications (Does everyone understand the alternatives applicable to the contracts as awarded?):

B. Initiating Construction (Notice to Proceed):
C. Completion Time for Contract (Does everyone understand contract requirements and methods of Computing?):
D. Liquidation Damages:
E. Request for Extension of Contract Time:
F. Procedures for Making Partial Payments:
G. Guarantee on Completed Work (Materials, Installed Equipment, Workmanship, etc.):
H. Other Requirements of the Contract and Specifications which Deserve Special Discussions by All Parties:
7. Contractor's Schedule: A. Analyze Work Schedule in Sufficient Detail to Enable the Engineer to Plan his Operations (Consideration must be given to needs of the Owner and the planned operations of other contractors):
B. Equipment to be Used by Contractor:
C. Contractor's Plans for Delivering Materials to Project Site (Protection and Storage of Materials):
8. Subcontracts (Review and approval of proposed subcontractors and their work schedules):

9. Status of Materials Furnished by Owner:
   A. Schedule for Future Deliveries:

   B. Procedures to be Adopted by Contractor in Accounting for and Storing Such materials:

10. Change Orders (Detailed explanation of procedure to be followed and clearance which must be obtained before changes are implemented):

11. Staking of Work (Clearly define responsibilities of Engineer and Contractor, Line and Grade must be furnished by Engineer.)

12. Project Inspection:
    A. Functions of the Engineer, Including Records and Reports:

    B. Responsibilities of Owner:

    C. Safety and Sanitary Regulations:

13. Final Acceptance of Work (Include requirements for tests and clean-up of project site):

14. Labor Requirements:
    A. Equal Opportunity Requirements:

    B. Davis-Bacon Act:

    C. Other Federal Requirements:

    D. State and Local Requirements:

E. Union Agreements:
F. Reports Required:
15. Equal Employment Provisions of Contract:
16. Rights-of-Way and Easements: A. Explain any Portion of Project not available to Contractor:
B. Contractor's Responsibilities During Work Covered by Contracts:
C. Coordination with Railroads, Highway Departments and other Organizations:
17. Placement of Project Signs and Posters:
18. Handling Disputes:

NOTED AND CONCURRED WITH, But understood not to be a modification of any existing contracts or agreements:

(Contractor Representative) (Contractor Representative)

(Architect/Engineer Representative) (Architect/Engineer Representative)

(Owner Representative) (Owner Representative)

APPENDIX 4B
SAMPLE PRE-AWARD CONFERENCE CHECKLIST

PRE-AWARD CONFERENCE
CHECKLIST FOR CONSTRUCTION PROJECTS
AND INSTALLATIONS OF MACHINERY AND EQUIPMENT

Date ____________________________

Job No.__________________________ Trade ______________________

Contractor _______________________ Representatives in attendance

____

____

____

____

Construction Manager Representatives in attendance

____

____

____

____

1. Base bid $________________ Includes addenda Nos ______________
2. Is price firm and inclusive of all taxes? ______________________
3. Has contractor visited the site? ______________________________
4. Questions on plans for specifications. (none_____) (see attached sheet)
5. Are there any exceptions contained in the bid which pertain to availability of materials, accessibility of site, etc? ______________________

   ____

   ____
6. Delivery or completion schedule: (including submission of shop drawings)

   ____

   ____

   ____

   ____
7. Does bid include all overtime necessary to man the job, meet the schedule, make all cut-ins and/or cut-overs and avoid interference with other contractors operations? (yes_____) (no_____). If "no," describe qualifications, use additional sheet if necessary.
8. Authorized overtime to be reimbursed at net cost of premium portion of pay plus applicable insurance and taxes, without mark-up for overhead and profit. (yes_____) (no_____)

9. Fees for additional work: Contractor's forces ____________
Subcontractors ____________
Allowances ____________

10. Unit prices ____________
11. Composite labor rates ____________
12. Alternates ____________
13. MBE/WBE participation ____________
14. EEO requirements ____________
15. What are payment terms agreed upon? ____________

____

16. Review acceptability of construction managers general conditions ____________

____

____

17. Did you review required insurance coverage; review fire, safety and security regulations ____________
18. Make specific reference to guarantee/warranty period ____________
19. Did you inform contractor to have inbound shipments identified and marked for his account and shipped prepaid: ____________
20. Contractor's proposed field organization: ____________

____

____

21. Detail of understanding as to what facilities, utilities, tools and equipment may be used, e.g. cranes, power, water, air, fork trucks, storage space inside or outside, field office on premises, etc. ____________

____

____

22. Proposed subcontractors and/or suppliers ____________

____

____

____

____

23. Items to be confirmed by contractor ____________

____

____

____

____

____

Signature of contractor: ____________________

Date: ____________________

Witnessed: ____________________

____

Signature of preparer ____________________

Date ____________________

CHAPTER 5

Contract Award Documentation

BID TO CONTRACT

§ 5.1 We have discussed the major ingredients necessary for the constitution of the contract. One of those ingredients is a meeting of the minds. However, many times the parties to a contract have good intentions, but in reality there is no meeting of the minds. An example of this can be seen when a public agency, anticipating a bond issue to be passed by the voters allocating funds for a particular project, will proceed with preliminary negotiations. If these preliminary negotiations mature into a final signed contract and the funds are never allocated to that department for that specific project, then the other party may not receive remuneration. Although the contracting party can sue the agency, in past cases the courts have not granted damages when there was no money. Obviously, from this experience, the burden remains with the contracting parties to determine, beyond a shadow of a doubt, that the funds necessary to complete their contractual duty are available.

This problem can also rear its ugly head in the change order arena. If sufficient funds do not remain in the contingency fund for change orders, and a change order is approved over and above the remaining amount, then the contractor will not receive remuneration. Again, although this appears to be unfair, the burden is on the contractor to determine the availability of funds in the agency's allocation.

Another concern regarding entering into a contract with a public agency is the capacity of the contracting party. Not only must the funds be available, but the party must have legislative authority to contract for the goods or services in question; *see, e.g., Atkin v. Kansas,* 191 U.S. 207. The contractor must look beyond the agency itself to the individual who will sign the contract. Governmental agencies are not held responsible for, nor are they bound to, agreements signed by unauthorized personnel.

It is also critical to note that organizations which are not properly incorporated and/or licensed in a particular state do not have the right to enter into contracts, even for private work, in that state. Although many organizations do enter into contracts and perform the work required by the contract, the validity of a contract will not come into question until some dispute is brought to court. Once the parties come before the bench, they must identify themselves and establish their legal right to conduct a particular line of business in that state. Not only must the contracting party be licensed or incorporated in the state in which it is performing work, but that status must have been achieved prior to the signing of the contract. Although the reading of the statutes refers to contractors, it is always assumed that a subcontractor is performing as a contractor in the eyes of the court—that is, in relation to statutes establishing the validity of its existence in that state.

In the case of *Sanjay Inc. v. Duncan Construction Co. Inc.*, 445 So. 2d 876 (Ala. 1983), Duncan contracted with Sanjay, as a sub, to furnish certain work for the project. At that time, Sanjay was a foreign corporation not qualified to do business in Alabama. However, eight months later, Sanjay did qualify to do work in Alabama. Subsequently, Sanjay filed a mechanic's lien against Duncan for $270,000. The appellate court said that a foreign corporation cannot enforce a contract if the foreign corporation had failed to qualify to do business in Alabama on or before the date the contract was made. In a typical situation where a foreign corporation is brought before the bench, the contract will be rendered null and void and any profit made from that contract will have to be returned.

In the North Carolina case of *Duke University v. American Arbitration Association*, 306 S.E.2d 584 (N.C. 1983), the construction manager, Turner Construction Co., employed Brunemann & Sons on a $1.5 million contract to fabricate and erect stucco wall panels. During the three years that Brunemann worked on the hospital project, it did not hold a general contractor's license for the state of North Carolina. When the university refused to pay Brunemann the $1 million extra work claim, Brunemann filed for arbitration. Duke asked a North Carolina trial court to stop the arbitration proceedings, claiming that Brunemann was barred because he did not have a general contractor's license. Both the North Carolina trial court and the court of appeals concluded that Brunemann was not a general contractor, and consequently did not have to have a license. They stated that in order to be a general contractor, Brunemann had to exercise control over the construction project as a whole, and over the work of the other contractors on the job. Because he had control only over his own work force, he could not be considered to be a general contractor. As one can see from the above cases, it is risky to do business in a state where one is not officially permitted to do so.

Once it is established that a business is contracting with an agency that has legislative authority, that the individual is authorized by the agency, and that the funds have been properly allocated, can the contract be signed? No. Now it is time to read the contents of that agreement form, the general conditions, and any other documents which bind the parties to the contract. These cautions, of course, are also true in the private sector, particularly regarding the availability of funds, the capacity of a party to contract, and the composition of contract documents.

In the general conditions, many clauses either shift a risk or disclaim it completely. If the risk falls on the contractor as a result of these two types of clauses, the contractor will want to make sure that he or she is willing to take on the risk, or to take measures to protect against it, or to shift it to another party.

INDEMNIFICATION CLAUSES

§ 5.2 A typical disclaimer clause in most construction contracts is the "infamous" indemnification or hold-harmless clause. In this clause, the design professional, in harmony with the owner, requires the contractor to indemnify or hold-harmless both the designer and the owner "against claims, damages, losses and expenses . . . arising out of or resulting from the performance of the Work . . . [which] is caused in whole or in part by negligent acts or omissions of the Contractor." A.I.A. Document A201, "General Conditions of the Contract for Construction," Article 3.18.1 (1987). What this clause simply does is relieve the owner and the designer from any liability for loss or damage due to the neglect of the contractor or any of the contractor's subs.

The indemnification clause is an explicit disclaimer clause familiar to all those involved in contract administration in the construction industry. However, many implied disclaimer clauses contained in contract writings are intended to serve the same purpose. Examples of such clauses are those contained in the A.I.A. Document A201, such as:

> Art. 4.2.3 The Architect will not have control over or charge of and will not be responsible for construction means, methods, techniques, sequences or procedures, or for safety precautions and programs in connection with the Work [and] will not be responsible for the Contractor's failure to carry out the Work in accordance with the Contract Documents.

> Art. 4.2.12 Interpretations and decisions of the Architect will be consistent with the intent of and reasonably inferrable from the Contract Documents . . . the architect will endeavor to secure faithful performance by both Owner and Contractor, will not show partiality to either and will not be liable for results of interpretations or decisions so rendered in good faith.

In the above clauses, a position has been taken by the architect in the contract agreement between the owner and the contractor not to be responsible for a particular condition. In effect, this clause is a disclaimer clause in that the architect states that he or she "disclaims" any responsibility for the loss or damage which may emanate from that condition.

JUSTICE IN THE BALANCE: ANTI-INDEMNIFICATION STATUTES

§ 5.3 The indemnification clause has long been relied upon by both the design professional and the owner as a shield or form of protection in cases where either personal or property damages were suffered during the construction period. Due to the unfairness of many of these clauses which protect one party to the contract but do not afford a similar and equal protection to the other party, many states have legislated anti-indemnification statutes. These statutes render such clauses unenforceable when the protected party may be a contributor to the negligence.

A key phrase contained in such statutes holds unenforceable clauses which "indemnify or hold harmless another person from that person's own negligence." Before one can analyze the impact of this phrase, one must understand the meanings of terms "indemnify" and "negligence." *Black's Law Dictionary* defines "indemnify" as "to secure against loss or damage . . . of an anticipated loss." It defines "negligence" as "that legal delinquency which results whenever a man fails to exhibit the care which he ought to exhibit, whether it be slight, ordinary or great; the omission to do something which a reasonable or prudent man . . . would do, or the doing of something which a reasonable or prudent man would not do."

Looking again at the key phrase contained in the anti-indemnification statutes, one discovers that it pertains to indemnifying one against his or her own negligence. It does not make reference to an indemnification clause which protects a person against the negligence of another person.

STANDARD CONTRACT CONDITIONS

§ 5.4 In looking at A.I.A. General Conditions Article 3.18, "Indemnification," the first seven words of the article are a severe warning to the user as to the application of this clause. They read "to the fullest extent permitted by law." The first thing that a user should do is inquire, preferably of the attorney retained by the owner, as to the validity of this clause for a proposed project for a specific location. If the project is proposed for a state where there is an anti-indemnification law, then the article must be scrutinized, preferably by the owner's attorney, to determine its enforceability. The A.I.A. article continues, "provided that any such claim . . . caused in whole or in part by negligent acts or omissions of the Contractor, Subcontractor . . . regardless of whether or not [it] is caused in part by a party indemnified hereunder." The wording "regardless of whether or not it is caused in part by a party indemnified hereunder" is suspect and leans toward the possibility of indemnifying one against his or her own negligence. This, as we have seen, is against public policy and is wholly unenforceable.

The fact that the A.I.A. General Conditions contain an indemnification clause does not indicate that it will be enforceable in a specific state. On the contrary, the A.I.A. strongly recommends that the user scrutinize the application of these documents. The A.I.A. has published guides for this purpose. The user of these documents should recommend to the owner that the owner secure legal counsel and that the owner's attorney review these documents and modify them accordingly.

ARBITRATION CLAUSES

§ 5.5 Most construction contracts also contain an arbitration clause requiring the parties of the contract to settle their disputes by means of arbitration as opposed to litigation. The arbitration that this clause refers to is that procedure established by the American Arbitration Association for the construction industry. The A.I.A. General Conditions contain such a clause and bind the contractor and the owner to dispute resolution via the arbitration process. However, in several states, such a clause is not recognized as enforceable, and therefore the parties do not have to go to arbitration. The logic behind this thinking is that the parties should first know what the dispute is about before agreeing to arbitrate. When the clause is contained in the writing long before the dispute arises, the parties have no chance at determining whether they want to arbitrate that particular dispute.

Figure 5-1 helps simplify the presentation of this matter. States not recognizing the arbitration clause in the writing of the contract do not require the parties to arbitrate at the time of the dispute. However, at the time of the dispute, if the parties desire to go to arbitration, they may do so. In the other states where the clause is recognized, the parties must arbitrate at the time of the dispute, unless both parties agree not to arbitrate at that time. Whether a state recognizes the arbitration clause or not, if interstate commerce is involved in the dispute, it must be arbitrated according to the Federal Arbitration Act.

Another critical fact about arbitration is that once a decision is reached, that decision is final. Even if both parties disagree with the resolution, the decision stands. Unlike the legal process in the courtroom, the arbitration decision cannot be appealed to a higher board nor can a party request that the matter be shifted over into litigation. The courts recognize the arbitration process and uphold its rules and regulations. Arbitration cases may be brought to court, not to be reheard and decided again, but to render the decision enforceable by law. The arbitration board has no means to enforce its decisions, but the court does.

The only time that an arbitration decision does not stand is when there is fraud evident among members of the arbitration board and/or on the part of one of the parties to the dispute. If fraud is determined to have existed, then the court will vacate the arbitration decision and require a new arbitration hearing. Fraud is determined by showing a partiality of one of the arbitration board members toward one of the disputing parties. The degree of partiality substantiating vacation of the decision is one whereby a substantial business transaction was conducted sometime prior to the hearing. Arbitration clauses also flow down to subcontractors and follow in the same line of thinking as they do in the prime contract.

A new wrinkle occurred in the case of *R.W. Roberts Construction Co. Inc. v. St. John's River Water Management District*, 423 So. 2d 630 (Fla. 1982), when the arbitration

Arbitration

AGREED TO BY BOTH PARTIES

	TIME OF SIGNING	TIME OF DISPUTE
11 States	Agreement contained in writing	Unenforced unless agreement by both at this time
Other States	Agreement contained in writing (Texas – special note)	Must arbitrate unless both disagree at this time

FEDERAL ARBITRATION ACT (Interstate Commerce)

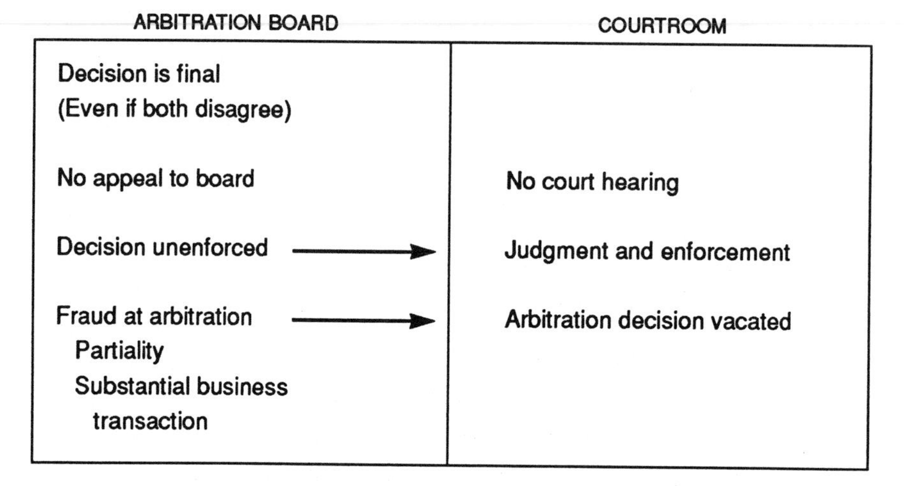

Figure 5-1

clause in the subcontract stated that "all claims by subcontractor against general contractor . . . which involve this subcontract or the project shall be submitted to arbitration in the same manner as provided in the general contract." MacDonald (subcontractor hired by Roberts) filed a complaint against Roberts seeking payment of $23,000 due under the performance and payment bond. Three weeks later, Roberts served MacDonald with the demand for arbitration. The court denied Roberts' motion to compel arbitration because the arbitration clause "lacked mutuality." The clause, as written, did not require disputes of the contractor to be arbitrated as it did the disputes of the subcontractor.

We could wave a red flag at this point and say that the contract should have been written by an attorney who was knowledgeable in this area and would have avoided this concern. Theoretically this is correct, but in practice most contracts, even those written by attorneys, have nebulous clauses which, until tested in court, are considered valid and enforceable. Certainly an attorney should be employed in writing the contract, especially the legal portion, and if that is not possible, then the least that can be done is to have the contract agreement form reviewed by an attorney prior to executing it.

A major concern regarding both the indemnification clause and the arbitration clause is the reliance upon these clauses as protection or means of settling disputes. An owner, for instance, may rely heavily on the indemnification clause as protection against suits by the contractor and the contractor's subs, only to find that the courts will not permit use of it because it's against state law.

INCORPORATION OF SCHEDULES

§ 5.6 A major means of documenting activity on a construction project is by use of a schedule. This schedule can be in the form of a simple bar chart or line diagram or it can be a sophisticated critical path method schedule. Either way, the major requirement of the user is official incorporation of the schedule as part of the contract documents. Once this is accomplished, the next steps are to periodically update it and to distribute copies to all parties involved in performing the activities designated on the schedule.

The *Kiewit* Case

§ 5.7 A case whose focus is on schedules is the landmark case of *Peter Kiewit Sons Co. v. Iowa Southern Utilities*, 355 F. Supp. 376 (D. Iowa 1973). Black & Veatch were the engineers from Kansas City, Missouri, who designed the project and, in the general conditions, called themselves project managers. The project had six multiple prime contracts. The contractors, by the nature of the operation, came on the site in phases. Peter Kiewit was general contractor, so obviously he was the first on the job. The job proceeded with Kiewit doing mass excavation and concrete foundations. The next prime to come in was the structural steel contractor. He did not make it on time. When he finally did get his steel in, it was improperly prepared and punched so that much of that steel had to be returned for refabrication. When the refabricated steel came back, his workers started to erect it. They had an accident and the steel fell, damaging other steel and the concrete. In addition there were strikes, bad weather, mud slides, floods, and other disasters.

The contract had a "no damages for delay" clause. It said, in essence, that the owner would provide an extension of time through the project manager, Black & Veatch, but that the owner did not have to pay any damages for the period of time extended. One of the key issues of the whole contract was the schedule. Actually, there were two schedules involved. One was prepared by Black & Veatch, and it was called a construction schedule. The second, a detailed schedule, was to be prepared by the contractors.

Of the six prime contracts, the boiler and turbine contracts were purchase orders and the other four were construction activities. Three of those contracts were awarded prior to the issuance of the schedule—the structural steel contract and the two purchase orders. The other three—general, mechanical, and electrical contracts—had the schedule issued with the bids. The structural steel contractor was given one critical date, the date when sufficient steel should be erected to hoist the drum of the boiler, at the time of the signing of the contract.

Black & Veatch prepared a construction schedule. It then asked the general, mechanical, electrical, and structural contractors involved in the construction process to submit, within 30 days after the award of contract, a detailed schedule.

The special conditions stated what both schedules represented, what the detailed schedule did, and what the construction schedule accomplished. It required the detailed schedule to comply with the intermediate dates on the construction schedule. The purpose of the construction schedule was to coordinate related work under the various construction and erection contracts to meet completion dates for certain phases of the work.

In his claim Peter Kiewit alleged negligence against both Black & Veatch and the owner in the following areas:

1. Duty to perform with the highest standard of the engineering profession. Black & Veatch said in its contract with the owner that it would perform in accordance with the highest standards of the engineering profession.
2. Preparation of plans, specs, and schedules in a manner that would allow orderly and efficient construction in accordance with the customs of the trade. Kiewit asserted that the engineer was negligent in the preparation of schedules.
3. Provision of adequate and competent personnel to administer the contract documents, which included the schedule, during the construction.
4. Provision of reasonable time extensions to contractors.
5. Delays to the general contractor as a result of unlawful administration of other separate contractors.

Kiewit also claimed that the owner breached in the contract by the owner's deviation from actual intermediate construction dates on the schedule and delays caused by the active interference of the engineer and the owner which negated the "no damage" clause.

In reviewing the written contract, the court looked at all the clauses contained therein and considered others which were absent. The court said that there was no express provision in the written contract that the intermediate dates were absolutely binding on the owner. Therefore, the "no damages for delay" clause was not negated and Kiewit would not receive any money for the extension of time.

This was a 1973 case. That's when the court made its final decision. The job was completed in 1967. This indicates that the documents were probably prepared about 1965. The case talks about Black & Veatch, an engineering organization, preparing a "construction schedule." Today we use different terminology for those two schedules. The first, the one prepared by the construction manager and issued with the bid documents, is called a "milestone" schedule. A milestone schedule does not call out the particular activities, but only "milestones" or critical dates.

In the *Kiewit* case, even though the schedule was titled differently, it was, in fact, a milestone schedule. The critical dates served only as a guide. According to the court's interpretation, the milestone schedule was to be used by the contractor as a guide to form the "construction schedule." This schedule is for coordination purposes only in order to meet the "critical" intermediate dates. It does not tell the contractors how to meet the dates or what to do in between the dates. Now the court did not narrow its review of the contract to just the schedule; it also looked at the clause in the contract writing which said the purpose of the schedule was "to coordinate related work." There had to be a schedule because two huge pieces of equipment were coming in which had to be coordinated with the construction process.

The clause continued, "under the various construction and erection contracts to meet completion dates for certain phases of the work." The court said that the schedule issued was just a milestone chart to give an idea of when various contractors needed to be on line. Since there was no "express" clause stating that the owner was bound to these dates, the contractor could not claim against the owner.

This case, with its method of issuing schedules, is mind boggling, especially since there were three contracts awarded even before the schedules came into existence. Then the engineers directed the structural contractor, who was one of the prime causes of the delay, to make up a detailed schedule based on the construction schedule that was never officially issued to him. They probably slipped him one, but not officially. Black & Veatch was not in trouble in this case because of a couple of things in the written contract. One is the wording that says that this milestone schedule is for certain stated purposes. The other is that there was no express clause that would bind the owner to the intermediate construction dates in the schedule.

Updating the Schedule

§ 5.8 As indicated previously, once the schedule is incorporated into the contract documents, it is important that it be updated and that the update be distributed to all those affected by the schedule. In order to update a schedule properly, the schedule itself must be monitored consistently to ascertain compliance by all those performing work on the project. Activities and dates which are indicated as critical to the project must be maintained, and if those dates are missed, notice must be given immediately to those responsible.

Typical contract language reads that all time limits stated in the contract, not simply the completion date of the project, are of the essence. If notice is not given to the parties causing the delay in meeting intermediate critical dates, then the right to claim damages when the project is not completed on time may be waived. Waivers are discussed elsewhere in this text but one should be aware at this point that a requirement of the contract writing can be waived by the action or inaction of the parties in carrying out those requirements.

Another benefit of monitoring the progress schedule is the effect on the morale of the labor force. Many projects suffer a loss of time and money because of improper scheduling of activities causing delays and disruptions in the daily work routine. The majority of the work force is proud of its work and looks to the time when each worker can say, "I built that project." That pride and motivation is eroded when confusion reigns due to a lack of proper scheduling. When confusion enters, disruptions and delays occur and morale dwindles. The best way to maintain the morale of the work force is to give workers short-term goals to reach and incentives in the process of pursuing each goal.

In many instances, a schedule is too sophisticated to be understood and carried out by the typical superintendent or foreman in the field. This schedule must then be translated into language which is understood by these individuals and presented in packages which are not beyond their scope of achievement. An example would be to take a particular activity from a critical path method schedule and lay it out on a bar chart or line diagram over a period of two or three weeks. This process can be repeated for every trade, with some indication of other trade activity that would parallel or possibly conflict with this particular activity. In so doing, the leader of that trade group has an opportunity to see exactly what will be happening over the next two or three weeks and what other trades will be paralleling that group's activity. These bar charts or line diagrams can also be used for monitoring the progress of the work, motivating the workers in a spirit of competition, and forecasting trouble areas before they become problems.

It is also worth repeating that these charts and diagrams become invaluable documentation for claim avoidance. On a project in New Jersey, as construction manager, we were able to avoid a $400,000 claim by presenting documentation on the contractor's activity over the previous several months, establishing him, and not the construction manager, as the cause of the problem.

The Schedule as Official Documentation

§ 5.9 Schedules must become an official part of the contract documents in order for a party to the contract to use them in claims preparation. If the schedule is merely a document produced by the contractor or a sub for personal benefit, then it cannot be used as an instrument against another party. If the same schedule is submitted to the architect for approval and the architect and/or the owner officially approve it, use it, and rely upon it for directing action and rendering decisions, then it will be considered official. Again, if that same schedule is not submitted and therefore not approved, but used and relied upon by the architect and/or the owner in a recorded progress meeting, then the courts may consider it official documentation.

EXECUTING SUBCONTRACTS

§ 5.10 In executing a subcontract, all the rules of contract writing exist just as they do between the prime contractor and owner, and as with all contracts, the more comprehensive and specific the subcontract is, the less likely it is that disagreements will arise over the work to be performed. Most subcontracts are executed on standard forms published by professional organizations such as the A.I.A. or A.G.C. Should the parties decide to create their own form, then the following conditions should at least be considered in the composition:

1. define the undertaking of the subcontractor (scope of the work) to eliminate any question as to what is required;
2. anticipate areas of dispute to determine the rights of the parties; and
3. define available remedies, specifying the relief to which each party is entitled. The contract should balance the rights and responsibilities of the parties without favoring one party to the point that the other party cannot accept the contract terms.

Other considerations include mandatory clauses. These clauses are applicable to the contract via federal, state, or local law which have an effect on the contracting parties. One example is the requirement for licensing in the particular state where the project is being built. (See the case of *Sanjay Inc. v.*

Duncan Construction Co., 445 So. 2d 876 (Ala. 1983).) Other mandatory clauses may refer to minority quotas as dictated by federal regulations, accounting procedures resulting from federal loans or grants, and the availability of subcontractors' records for auditing purposes even after the project is completed.

An excellent practice in the composition of subcontracts is to incorporate the same requirements that exist in the contract between the prime and the owner. This is achieved with a flow-down clause similar to that given in the A.I.A. Form A401 (January 1972 edition) for subcontracts. The clause reads as follows:

> 11.1 The subcontractor shall be bound to the contractor by the terms of this agreement and of the contract documents between the owner and contractor, and shall assume towards the contractor all the obligations and responsibilities which the contractor, by those documents, assumes toward the owner and shall have the benefit of all rights, remedies and redress against the contractor which the contractor, by those documents, has against the owner, in so far as applicable to this subcontract, provided that where any provision of contract documents between the owner and the contractor is inconsistent with any provision of this agreement, this agreement shall govern.

An area of major concern is the time of payment to a subcontractor based on the terms and conditions of the subcontract. Unless otherwise specifically noted in the contract, a subcontractor has a right to payment even though the prime contractor did not receive payment from the owner. In order to prevent such a situation, a "condition precedent" clause regarding payment must be inserted in the subcontract. This clause will establish that the subcontractor payment is dependent upon payment by the owner to the prime.

When a subcontractor is not paid by a prime, the sub has two avenues of pursuit regarding remedies and relief. The first is to claim against the prime's payment bond; the second is to lien against the owner's property. When a subcontractor claims against the prime's payment bond, the surety company will be required to pay the subcontractor if the sub can demonstrate the right to payment. The surety company is liable only to the degree that the prime contractor is liable. If the fault for the subcontractor's nonpayment lies elsewhere, then the surety company may not be required to pay. When pursuing a lien against the owner's property, one must be aware of local lien laws which may prevent the action.

In some states, one must be previously registered (prior to project beginning) before making a claim. The lienor must have substantially performed in order to be entitled to a lien. Who can lien is also governed by applicable statutes, which usually permit the prime, the sub, and material vendors to file such a lien. Some construction contracts require that the prime contractor waive the rights to lien. In some states where this requirement exists, the prime not only waives the contractor's right but all the subcontractors' rights. In the state of New York, the subcontractor's rights are subrogated and he or she can get no more than the unpaid balance in the hands of the owner that is due to the prime contractor. In Pennsylvania, a subcontractor's rights are direct and the subcontractor may be paid up to the limit of the property regardless of the unpaid balance due to the prime. Both prime contractors and subcontractors should be reluctant to waive their lien rights. Subcontractors especially should be alert to the possibility of a prime waiving the right to a lien along with the rights of all the subcontractors. Once again it is recommended that an attorney review contract agreement forms in light of the statutory requirements regarding liens and lien waivers.

CHAPTER 6

Legal Aspects of Construction Documentation

ORAL DIRECTIVES VS. WRITTEN DOCUMENTATION

§ 6.1 In the daily routine of conducting business on a construction site, many directives are conveyed orally. There are several dangers in directing activities orally, one being that the oral directive may not be clearly understood and therefore not accurately carried out. To guard against such errors, all oral directives should be substantiated in writing to help clarify the information conveyed. In many instances, it would be beneficial to also request a response to the written directive so that there is no misunderstanding on the part of either party.

Written directives are not only useful for clarification but also for confirmation and creation of a permanent record. A telephone log can be very beneficial for not only recording all telephone communications but also for confirming the communication via a written letter. This establishes two goals: one to confirm in writing what was spoken over the phone, and the other to have a permanent record of that communication. In the process of confirming the oral communication with the other party via a letter, the parties also can clarify any misunderstanding conveyed over the phone.

All written communiques such as letters, minutes of meetings, and field reports automatically become official documents pertaining to the project. These written communiques become supporting documents to the original contract documents in that they can establish where the original contract documents were carried out accordingly.

Written documentation is a form of certification which, according to the meaning of certification, establishes a historical record. An even more accurate certification is to issue a writing and request the other party to sign and return a copy for the record.

PRECEDENTIAL ORDER

§ 6.2 In addition to the precedential order of the contract documents for a construction project, there is also a precedential order for any contract based on general legal principles. This order from the least important to the most important is as follows:

Printed Form (Standard Universal Language)

§ 6.3 The general conditions are a typical example of a printed form containing standard universal language used in almost every construction contract. These forms are prepared by professional organizations such as the American Institute of Architects (A.I.A.), Association of General Contractors (A.G.C.) or National Society of Professional Engineers (N.S.P.E.). Since these documents are prepared for the use of every possible size and type of construction project, the language truly is "general." The major point is not the general language itself but the fact that it is a printed form. When compared to other documents, the standard printed form carries the least weight.

Typed Form (Specific Project Language)

§ 6.4 A typical example of a typed form changing, or having precedence over, a printed form is the supplemental conditions of a construction contract. In the case of *Otis Elevator Co. v. Don Stodola's Well Drilling Co.,* 372 N.W.2d 77 (Minn. App. 1985), Stodola submitted a verbal bid followed by a written proposal. The document contained a typewritten clause that indemnified Stodola from any damage to the building which may result from the installation. When Otis issued a contract to Stodola, it was on a printed form which also had Stodola indemnify Otis for any property damage during performance of the work. The court held that the clause on Stodola's typewritten proposal took precedence over the printed indemnity clause in Otis' purchase order form.

Handwritten Form (Specific Activity)

§ 6.5 When two parties come together to sign a preprinted or typewritten contract, they may not agree to all the terms contained in it. Should the change of one of those terms be relatively simple, then the parties can insert a handwritten note, initial it, and date it. This handwritten note will take precedence over the previous requirement. This is also true in the construction industry during field activities. When an architect makes an effort to be expedient to maintain the project schedule and issues a directive on a handwritten note, that note takes precedence over the previous requirement whether it was a graphic representation, typewritten, or printed form.

Oral Directive (Specific Activity)

§ 6.6 An oral directive, like a handwritten directive, issued in the field, will take precedence over any previous requirement of the original contract documents. In the case of *Tamarac Development Co. v. Delamater, Freund & Assoc., P.A.*, 675 P.2d 361 (Kan. 1984), the architect orally agreed to supervise the grading construction and ensure its accuracy. When the grading was found to be inaccurate, the court held the architect liable for an implied warranty to inspect and supervise the grading committed to in the oral agreement.

TIME LIMITATIONS

§ 6.7 In some instances, no matter how accurate the documentation, or how serious the violation, the claim for damages will be rendered void if the documentation of that claim is not submitted in a timely manner. In the Nebraska case of *Omaha National Bank v. Continental Western Corporation,* 274 N.W.2d 867 (Neb. 1979), the court would not permit an architect to file a mechanic's lien against the property because the claim was filed too late. It was undisputed that Bennett, the architect, had performed appropriate work to justify the filing of a lien against the property. However, the mechanic's lien laws set certain limits for the time within which the claim can be filed. Time limits for mechanic's liens and similar avenues of claim are strictly enforced by the courts.

DOCUMENTING PROCEDURES

§ 6.8 Due to the repetitious nature of much of the documentation done during the construction process, many forms have been created to help facilitate this documentation. What these forms essentially do is take the repetitive elements of a process and record them on a standard form to be used over and over again. The variables included on the form are inserted in blank spaces provided in the form.

Examples of these forms include the change order form issued by the architect on behalf of the owner to direct the contractor to perform work over and above that required by the original contract documents. As we have stated earlier, the change orders affect the time and/or cost of the project. Therefore, provisions are made on the form for these items to be incorporated as variables. Some of the repetitive language of the change order form might include the directive to proceed with work immediately upon receipt and to incorporate payments for this work in the regular monthly requisition.

Another form that helps to facilitate the documentation is the request for payment form. This form may require the contractor to have the document notarized before submission to the architect. Other standard forms for payment requests may require an itemized accounting of every trade (subcontractor and/or supplier), giving the last amount paid, the total amount of the subcontract, and the amount due this period. These documents are expressed in a standard form with the repetitious elements (names and activities of the various trades) printed and the variables (the amounts due, etc.) typed in the blank spaces.

Many of these forms can be obtained from professional organizations such as the A.I.A., the A.G.C., and the Engineers Joint Contract Document Committee (E.J.C.D.C.). These forms are recommended for use by those participating in the construction process since they have been prepared by professionals familiar with the requirements for that particular activity. However, one should be cautioned against the use of these forms without first scrutinizing them for validity on a particular project. Some forms may contain repetitious elements which are not required for a specific project and which would cause confusion for the other party in complying with the form's requirements.

On large projects covering a long time, it may be beneficial for an organization to create its own custom forms. These forms can include not only the repetitious requirements in the way of activities performed and recorded, but also regarding the distribution of the form. On the Philadelphia Stadium project, where we had more than 12 prime contractors involved, the transmittal form was designed to incorporate all of the prime contractors on the distribution list at the bottom of the form. This facilitated the processing of the form with a simple check before the individual organization requiring copies. When designing these custom forms, one must be aware of the fact that they must be complete and easy to use and understood by the parties on either end of the transaction. This is accomplished by organizing the form in such a way that the constants clearly dictate how and where the variables are to be inserted. Efforts should also be made to minimize the number of variables and to maximize the number of repetitious requirements.

MAINTENANCE AND RECORDING SYSTEMS

§ 6.9 In addition to using forms for facilitating the transmittal of information, one can also readily use systems for maintaining and recording current activities. One such device is the use of logs for recording the activity or processing involved in transmitting shop drawings, change orders, and requests for payment to the different organizations. An example of the use of logs is in the tracking of change orders. Unless a change order is properly tracked by means of a log, it can easily be delayed in the processing, causing an additional increase in cost to the project as a result of having to remove original contract work installed during the "delinquent" period.

Charts are an excellent means of recording current activity in the field. One such activity might be the installation of windows on each individual floor of a high-rise building. The chart can be used to indicate the projected schedule as well as the actual progress. These charts are then recorded in the minutes of meetings or in the file for permanent record. They become an excellent tool for preparation of claims, should there be default by the subcontractor installing the windows. Another way to use bar charts is to list all the activities during the construction phase in sequential order and the location of those activities. These two elements will become constants on the chart and the variables will be the progress and the date that the meeting is held to review the chart. This will give a permanent record of where each trade was on a particular date.

"A picture is worth a thousand words" can easily be considered an understatement in a construction-related claim. Photographs of activities and conditions on a construction site are invaluable, especially if properly recorded, documented, and distributed. In addition to the typical progress photos, which are taken from strategic points at regular intervals, additional photos should be taken of specific concerns in areas which may become the subject of claims. A professional photographer should be used and the photos should be properly designated on the face of the photo with the project title, date of photo, and position of photo. If these photos are to be distributed, they should be more accurately identified with a cover letter and a reference to the areas of concern.

In distributing documentation, one must be aware of contractual relationships and must not distribute to those who are not party to a contract or who may be detrimental to one party's efforts. An example of a detrimental event would be to carbon copy a letter from the architect and owner to a subcontractor, which points out the discrepancies in the subcontractor's work. Typically, distribution should be made to those who are party to a contract, attendees at a meeting, and those affected by the contents of a communique. Distribution may also be to support organizations such as the surety company of the subcontractor, if the subcontractor shows signs of going under.

RECORDING PROCEDURES

§ 6.10 Probably the greatest event that occurred in the construction industry regarding record keeping and data retrieval is the introduction of the MASTERFORMAT system of section titles and numbers by the Construction Specification Institute (C.S.I.) and Construction Specifications Canada. What this system does is create a category for every possible material or piece of equipment used in the construction industry.

These categories are broken into 16 divisions, each titled for a major area of concern (such as concrete, masonry, metals, etc.). Under each division heading is a series of sections that deal with specific concerns (concrete block masonry, stone masonry, brick masonry, etc.). These division headings and specification section headings never change. They are always in the same order and one can rely on finding product data by looking under the specific division and section number designated by the system for that product. For instance, if a contractor were looking for information on the mortar mix for a concrete block wall, the contractor would look under Division 4, titled "Masonry," and under the section titled "Mortar and Masonry Grout."

Although this system is rigid, and it need be in order to maintain uniformity throughout the industry, there is flexibility within the section organization. This flexibility can be achieved by C.S.I.'s recommendation to use the sections for either "broadscope," "mediumscope," or "narrowscope" categories. A broadscope category would be one that includes several mediumscope and narrowscope topic headings; a narrowscope category includes only one section topic heading. An example would be to incorporate several section headings under one broadscope section titled "Curtain Walls." Under this heading would be the curtain wall itself, the glazing requirements, the caulking, anchoring devices, panels, and even louvers, if required by the system. In a narrowscope breakdown, each one of those headings would be a single section. The MASTERFORMAT broadscope section titles are set forth in Appendix 6A at the end of this chapter.

Several factors dictate whether a category should be broadscope, mediumscope, or narrowscope. One is the amount of information available for a particular section. If very little information is available, then several sections can be joined together to form one file or one section in the specification. Another factor would be the assigning of responsibility to a particular subcontractor. For instance, with the curtain wall system discussed above, all of those activities should be listed under one broadscope section so that if there is a failure in any one of those areas, one subcontractor can be held responsible. Imagine the chaos had every one of those requirements been installed by a separate subcontractor.

Another recording procedure, the chronological file, has been around for a long time, and is simply a method to record data based on the date it was documented. This is typical for daily correspondence, such as letters, transmittals, and minutes of meetings. Many organizations who use the chronological file use it in conjunction with other filing systems such as the topical system created by C.S.I.

Claims files are another recording procedure and are becoming more and more necessary for participants in the construction industry. A claims file simply takes data surrounding a particular concern, even from a very early date, and compiles it in one file. Should the concern develop into a valid claim, then all of the data pertaining to that claim will be chronologically filed in a specific claims file for easy access. As this file develops, efforts should be made to include not only the routine correspondence, but also additional data instituted by members of the organization, such as special photographs or minutes of special meetings or special field reports.

An action file (tickler file) is one which permits a user to have automatic reminders of daily functions which must be conducted to maintain the progress of a project. An example would be to direct a subcontractor to deliver products to the job site by a certain date or to employ additional people by a certain date and record this requirement in the daily tickler file. The file is broken down into the five working days of the week.

At the beginning of each day the user looks at the tickler notes contained in that particular day's file and then conducts the activities required by those notes.

NEED FOR CONSTRUCTION PHASE DOCUMENTATION

§ 6.11 While proper documentation in the bid and award phases are critical to securing and establishing a profitable contract, proper, accurate and consistent documentation during the construction phase could mean economic life or death to the contracting parties. Every member of the construction team must document every activity performed by team forces and those who would affect the operation. To do anything less could be the demise of a contractor's organization.

Material Storage

§ 6.12 A typical construction contract between an owner and a builder includes a clause which states that if a contractor is permitted to purchase material in advance and store it on or off site, then the contractor will be reimbursed for the amount, minus a percentage set in the contract writing. Before the owner pays, there should be some established procedures to ascertain ownership and proper protection. What are some of those procedures? What type of building? Is it a warehouse? What kind of warehouse? Who owns the building? Who has control over it? These are serious questions which must be asked before issuing payment, and even before writing the contract conditions. When the owner says that he or she will pay the contractor, the contract should stipulate under what conditions payment will be allowed. Some of the conditions are:

1. The material is to be marked indelibly for this project.
2. The percentage to be withheld must be designated.
3. The owner is to receive a bill of sale indicating ownership of the material or equipment.
4. The product will be stored in a bonded warehouse. What is a bonded warehouse? The archaic definition of

"bonded" comes from the old alcohol and tobacco industry. Materials would be put in a warehouse and the door sealed and the warehouse would be "bonded" for a period of time. That's not today's custom. What is probably meant by a "bonded" warehouse is one that is insurable.

5. The material and/or equipment itself must be insured. Why should one insure the building and the product? In New Jersey, a very prominent, nationally renowned manufacturer called us to tell us that our equipment was complete, stored in its warehouse, and ready for inspection. A representative of the company added that the company would transfer the required paper work, and then the company wanted payment. A state representative and (the construction manager) went out to certify the requisition. The representative took us through the building where the material was stored. Everything was properly identified for our project, stacked orderly on shelves, and wrapped for protection. Everything was in order. The paper work was in order. As far as I was concerned, I was ready to sign it off. The state representative asked the manufacturer, "What is this operation going on over here?" He said, "That's the way we inject the insulation into the molds." The state representative said, "I can't pay you." Our insurance may no longer be applicable if it is stored in a manufacturing plant.

In another example, one of my colleagues, Karl Zintl, president of a construction management organization in Chicago called Project Control, Inc., purchased walkie-talkie units with a base station, a copier, typewriters, calculators, and other equipment, and put them in a trailer on a construction site. When the building was substantially complete, he took all that equipment and put it in a room and secured the door. One night someone kicked the door in and stole everything. Karl put in a claim with his insurance company. The claims adjuster came, looked at the room, saw the door kicked in and said that he would have to deny the claim. The policy showed that all the equipment was in a trailer when it was insured. It was covered by the policy when it was in the trailer but not after it was moved to the building. The policy was explicit. All Karl had to do at the time of the move was to call the insurance company and tell them to change the policy to read "in the building." He didn't do it. He never even thought of it.

Insurance companies can catch a contractor on a technicality. One has to have the material insured, have the warehouse insured, make sure the location is a warehouse, make sure the paper work is in order, and make sure every piece is designated for the contractor's job. Is that a bizarre requirement? The manufacturer could be making equipment for some other job and charging the contractor. For example, there was an incident where a supplier asked an owner to pay for 400,000 lineal feet of 3/4-inch copper tubing stored in the supplier's warehouse. The owner sent a representative to make a visual inspection of the quantity, quality, and location where it was stored. The paper work was in order, so the owner paid. That same supplier went to another owner the next day and received payment in a similar fashion. One may say that the owner didn't really lose because the material will be used when the project is ready for it. Will it? What about the other owner who may need it at the same time? If it's not marked, don't pay for it.

Another concern is the amount of reimbursement for materials stored off site. In an experience in New Jersey, the precast concrete manufacturer would produce so many panels every month, store them in the yard, properly mark them, and have all paper work and everything else in order for inspection and subsequent reimbursement. We would go and observe it and pay the total request minus the 10 percent retainage. We did that month after month until one month they closed the doors of the plant.

The company declared bankruptcy. What does a contractor do when a company declares bankruptcy? One can't do a thing. Even though the contractor can prove that the material is his, not until the legal process is settled can he get his materials. In this case, I'm talking about enough panels to enclose a building that covered a city square, 10 stories high. That building sat for months with the wind blowing through it. Finally the legal matter was settled and the company told the contractor to come and get the panels. The contractor told the company that it had to deliver the panels as part of the contract. The attorney reaffirmed that the company was still out of business: "If you want these panels, and they are yours, you had better come and get them." It cost $80,000 to rent two cranes and flatbed trucks to handle and ship those panels to the job site. When we paid the contractor, we paid for materials, shipping, and handling. What did we really get? Materials. The contractor was the one buying it from the precast outfit. A contractor, who has built buildings around the world, a state representative, an architect, and the construction manager all overlooked the cost of shipping and handling.

A contractor should only pay 50 percent of the value of any material stored off site. This is a good policy to adopt and incorporate in a contract. (It must be in the written contract before it can be enforced.)

Shop Drawings and Samples

§ 6.13 Who approves shop drawings? How about the construction manager? A construction manager, who accepts responsibility, along with the architect, for approving shop drawings is then responsible. Is it possible that the construction manager is in an area where he or she doesn't belong? Historically, architects are responsible for the review of shop drawings and samples. It is inherent in their responsibility to see that the contractor is complying with the contract documents, and unless it is specifically written out of the contract, they are going to be held responsible for the review and approval of shop drawings.

The architect/engineer should review shop drawings and samples. It is the construction manager who monitors the processing of them and, in particular, the time frame. What is actually happening is that the construction manager is monitoring the architect for turnaround time. The owner should hold the construction manager responsible if the shop drawings are not processed back and forth on time, but should hold the architect responsible for the content and the review and approval.

Payment Certification

§ 6.14 Who signs the payment request? It is always signed by the architect. Why? Because the architect/engineer is the one who has to inform the owner that the contractor has been building according to the design and therefore should be paid. Should the construction manager also sign the request? On the Philadelphia Stadium, the construction manager was primarily the inspector and every month would be the first one to receive the requisition for payment from the contractors. Because the construction manager was the "eyes" of the management team, it was natural that the construction manager would know the quantities and quality of the work. The construction manager would review the request but was not required to sign it.

I recommend that the construction manager, who is representing the owner and is primarily concerned with the owner's budget, sign that request, and also monitor the processing of it, even to the point of monitoring the owner paying the contractor. That is part of the construction manager's responsibility. The owner owes that money for the work completed up to that point in time, and the construction manager needs to advise the owner accordingly.

Change Orders

§ 6.15 No owner wants a change order. Even though there is a contingency fund for such events, the owners would rather not have to deal with a change order. When a change order is processed, it must be signed by the owner because it may have an effect on the schedule, and will usually have an effect on the dollar amount. The architect signs the change order because the change has an effect on the design, and the architect is responsible for the design; therefore, no one is permitted to alter that design until the architect authorizes the change.

What does the construction manager have to do with a change order? Will the change affect the schedule? It might affect it seriously. Will it affect the budget? Definitely. Therefore, the construction manager should know about it and address these two major concerns.

Two hidden legal principles emanate from a change order that have a serious effect on the budget and schedule: acceleration costs and waivers.

Acceleration Costs

§ 6.16 Acceleration of a project is like acceleration of an automobile. To get it going faster, one accelerates or increases the rate of speed. Sometimes when one wants to get a project going faster, one accelerates it. There are times when an owner will require the contractor to do more work within a fixed period of time through a change order. In order to get it done, the contractor will have to accelerate the project by bringing in more crews or having existing crews work overtime. Sometimes, owners will put out change orders, increasing the amount of work but not giving an extension of time. One of the reasons for doing so is that the owner wants the job done on a particular date, and it is critical. The owner who does this runs the risk of having the contractor come back and claim acceleration charges.

An acceleration clause is not a typical clause in a construction contract. There won't be a clause in a contract that says the owner will pay acceleration costs. How does one secure acceleration costs? Acceleration is either "directed" or "constructive." This means that the owner will direct the contractor either to extend the period of time because of the extra work required, or to perform the extra work without an extension of time. The latter is called constructive acceleration. The validity of an acceleration clause requires that several factors be established:

1. There has to be an excusable delay.
2. An extension of time has to be properly requested, within a proper period of time.
3. The owner must fail or refuse to grant an extension.

4. Acceleration must be ordered or implied.
5. Acceleration must actually be performed.

The contractor must document the performance of the extra work by acceleration, either through working more people or working overtime. The work must be done according to the owner's schedule and not the contractor's. If it were the contractor's schedule, and that schedule never became part of the contract documents, the acceleration was for the contractor's benefit only, and the contractor would not have a right to claim acceleration charges. If it is the owner's schedule, then the contractor can claim acceleration charges.

How does the construction manager fit into this scenario? One of the major concerns of the construction manager is to bring the project to completion on the date scheduled. One way to accomplish this task is to deny the contractor any extension of time even if the contractor's request is legitimate. The construction manager may run the risk of not granting the extension with the hope that by the time the project is to be completed, the request will have faded into oblivion. Is this realistic? Would a construction manager purposely deny an extension of time even if it were legitimate? My experience, and that of many of my colleagues around the country, tells me that it is. On the Philadelphia Stadium project, every change order processed affected to varying degrees the amount of money requested. Not one change order was approved for an extension of time. In fact, at the outset of the project, the city's administrator ordered the entire project team not to grant any extensions. The opposite view is that the owner's interests are best served by timely grants of proper time extensions.

Waivers

§ 6.17 Again, as a result of change orders, the construction manager may be creating a waiver something without realizing it. Typically, a construction manager can waive job requirements either by an action or a lack of one. An example is when a construction manager walks a site and sees the contractor doing some work that the contractor should not be doing. It may be change order work which has not yet been documented or authorized with the signature of the owner. The construction manager is on the job site representing the owner, sees the contractor doing the work, and says nothing. The construction manager continues to conduct operations in this manner until one day there is a dispute. When the dispute goes to court, the judge may say, "You have waived your right by not telling the contractor that he was out of compliance." The construction manager may say that he or she was anticipating a change order for that work, which was not issued yet. The court will enforce the requirements of the contract, that no change order work be performed until the change order was validated by the owner's and architect's signatures. On the other hand, if the construction manager tries to tell the contractor that the construction manager will not pay for work on a subsequent change order because the work was started before the change order was signed and issued, the court will tell the construction manager that by his action, or in this instance lack of action, the construction manager has waived the right to require the contractor to wait for a signed change order to be issued before starting that change order work.

According to legal principles, several conditions will establish a waiver:

1. The owner was aware of the contractor performing some change to the work and did not object.
2. The item is of such magnitude that the changes could not be made without the owner's knowledge. For example, presume the original design did not have a canopy at the front entrance. All of a sudden there is a precast concrete canopy. The owner goes in and out of this area all the time. It is so monumental that the owner or architect has to be aware of it. The owner never says a word to anyone about it. Now the contractor requests payment for the canopy work and the owner refuses to pay. A dispute arises and the contractor says the architect told the contractor to install it. The owner says the contractor is supposed to have a change order signed

before doing change order work. The owner waived that requirement of the written contract by performance, or lack thereof, in the field.

3. The changes are necessary but not foreseen by the design professional. This may occur as a result of a piece of specified mechanical equipment which does not have a self-contained starter and the engineer did not specify a remote starter.
4. There is a subsequent oral agreement. The oral agreement waives the requirement of the written agreement.

In monitoring the change order process, there are many concerns—logging the activity, quick turnaround time, close technical scrutiny, etc. All are critical, but the two "hidden" concerns discussed above need to be brought to our awareness before we fall prey to their subtle grip.

Substantial Completion

§ 6.18 "The date of substantial completion is the day when the building, or part thereof, is complete to the extent that the building, or part thereof, can be used for the owner's intended purpose" is the definition given in most contracts. Now add these words, just for discussion: "whether the owner uses it or not." The date of substantial completion arrives when all the requirements defining that date come into reality. So on the date of substantial completion when the building, or part of it, is complete to the extent that the owner can, in fact, use it for its intended purpose, then the construction manager, the architect, the owner, and the contractor must come together and sign and date a certificate of substantial completion. That means certain items are transferred from contractor to owner. The warranties and guarantees start for all the equipment in that particular area or for the entire building. Insurance is transferred. Maintenance and operation of equipment becomes the responsibility of the owner. It is a critical date and it comes into existence whether the owner needs the building or not. If the owner should suggest that he or she really does not need the building, and wants the construction manager to hold the contractor off, the construction manager cannot do that. It is the owner's building. The construction manager has the responsibility of getting the owner to sign the certificate. Otherwise the construction manager could be held negligent for not performing a duty owed to the contractor.

If the owner wants to partially occupy the building, the construction manager may need to punch list that "area" and have it turned over under substantial completion. If there are any damages, the owner may have to pay for them. Heating, insurance, and air conditioning for that area are now the owner's responsibility. This procedure must be accurately handled and documented or there will be problems. When the owner is in under partial occupancy, there is always a hassle over the use of the systems and their efficiency. Remember, the building process is not complete. There may be construction dust seeping into the systems, rendering them less efficient.

Liquidated Damages

§ 6.19 The courts recognize that the owner, if he or she doesn't get the building on time, will be losing money as a result. For each day that the building is overdue, you calculate what the owner is losing on a daily basis, and that dollar amout is what is used as the liquidated damages amount. The courts have established conditions which determine the validity of a liquidated damages clause. In every jurisdiction, at least two conditions must be present and, in some jurisdictions, a third. These conditions are as follows:

1. There must be difficulty in determining actual damages. It must be hard to calculate the amount of money that will be lost on a daily basis. Instead of an arbitrary figure, the courts want the parties to figure out what the costs are to the owner for insurance, security, maintenance, and other charges in comparison to the owner's estimated income. If one would subtract the first from the second, the result is approximately the amount of liquidated damages.

2. The liquidated damages amount must be in line with actual damages. What is meant by actual damages? It is the amount that the owner will actually lose.
3. The owner must have suffered damages. This is a jurisdictional issue. In some jurisdictions, the condition is applicable and in others it is not. To "suffer damages" means that one must lose money or be injured in some way. When the day comes that the building is to be open but is not, the owner must show that he or she is losing money. If the owner is not losing money, then in some jurisdictions the owner cannot apply for and collect liquidated damages.

Is a penalty clause ever valid? Yes, it may be if there is a penalty with a bonus. This really is an incentive for reward as well as a penalty for default. The clause says that if the contractor does not get the job done on time, the contractor will be charged a set dollar amount for each day thereafter until completion. But, if the contractor finishes early, the two parties will split the difference. There is an incentive, and it is considered a just balance; therefore, the courts will permit it.

If, when a job is supposed to be complete, it is not, and the owner has a liquidated damages clause in the contract, the owner can hold monies from the contractor on a daily basis. The owner does not have to go to court to do so. If the amount is $1,000 a day, and the contractor goes 50 days over, the owner can withhold $50,000. The money is the owner's and it is still in the owner's account accruing interest. If the owner would have to go to court and prove actual damages, the owner would have to hire an attorney, who arranges for a hearing in court. It may be several more years before the court can hear the case. However, with liquidated damages, all other conditions being established, the owner can withhold money immediately without having to go to court. It should be the owner's choice, not the attorney's, as to whether to include a liquidated damages clause. As the representative or agent to the owner, the construction manager should make the owner aware of these differences. Remember, the construction manager is not permitted to act as an attorney because that would be practicing law without a license; but the construction manager can certainly discuss these concerns with the owner's attorney. It should not be the attorney who makes the final decision, but the owner.

The landmark case of *Bethlehem Steel Corp. v. City of Chicago*, 350 F.2d 649 (7th Cir. 1965), will show the benefit of incorporating a liquidated damages clause into a contract. The case involved a separate prime contract situation, and it was phased. The project was a high-rise freeway in Chicago. The first phase was the concrete foundation work. The second phase was Bethlehem Steel's, which required the company to put up the structural steel and steel deck. The third and fourth phases had to do with paving, rails, and lights. Bethlehem Steel had in its contract a liquidated damages clause which would go into effect on the day that Bethlehem was to complete its phase. The amount was $1,000 per day. As the date was approaching, Bethlehem realized it couldn't meet the date, so it put in for an extension of time. The city of Chicago extended the time. Bethlehem didn't make the extended date and went past it by 50 days. The city of Chicago withheld $50,000. The contractors for the third and fourth phases came on the job and finished their phases, not only on time, but early, so that the entire project was completed on time. Bethlehem Steel realized that the city of Chicago had its project done on time, and, therefore, did not lose any money. Bethlehem took the matter to court to retrieve its $50,000. The judge said that the company agreed to pay $1,000 per day for liquidated damages whether the city suffered actual damages or not.

In *Trans World Airlines, Inc. v. Travelers Indemnity Co.*, 262 F.2d 321 (8th Cir. 1959), TWA had a liquidated damages amount of $100 per day in its contract with a contractor. As the job extended past the completion date, the airline was losing thousands of dollars per day. TWA took the matter to court to get more damages from the contractor. The judge held TWA to the terms of the contract.

Time Is of the Essence

§ 6.20 What does this term mean? "Time is of the essence" means that the time factor is essential to the existence of the contract. So when "time is of the essence" is in a contract, time is what makes that contract. If the time factor were removed or breached, the contract would fall apart. In the A.I.A. General Conditions (A201, 1987 edition), "time is of the essence" is included. However, there is not a topic heading in bold letters stating "Time Is of the Essence." In Article 8, which is simply entitled "Time," in paragraph 8.2, the first line reads, "Time limits stated in the Contract Documents are of the essence of the Contract." It does not say the final completion date, it says simply "Time limits." If one were to put out a schedule in the form of a bar chart, an arrow diagram, or a critical path method schedule, and indicate milestone dates, all those milestone dates are critical to the life of that contract. If a contractor is not paying attention to those dates, the contractor may lose the contract.

What happens if the construction manager is not paying attention to that contractor's activities, the contractor misses that first date, and the construction manager doesn't say anything about it? Now the second critical date comes and the construction manager doesn't say anything about it. The third critical date comes and the situation is repeated. Finally, the final completion date of the job arrives and the job is not complete. The construction manager now withholds liquidated damages and the contractor is going to prove to that court that the time limits were not of the essence. When that first date came and passed by, nobody notified the contractor about the missed critical date. The second and third milestone dates came and passed, and again no one notified the contractor. Now it is the final completion date and the construction manager takes action. It is too late. The owner and construction manager have waived the right to act on those time limits. That right was waived when no one took action on those intermediate dates. Something should have been said and put in writing regarding those dates. Make sure missed dates are not a casual thing. Put out a formal directive. The construction manager cannot just say in the minutes of a progress meeting that the milestone date for enclosing the building was March 5, and the contractor did not meet it, and then go on to the next item.

Site Condition Changes

Subsurface Condition Changes

§ 6.21 A major area of concern regarding site changes is subsurface conditions. For years, both in my experience as an on-site representative for construction managers, contractors, and subcontractors and also as a lecturer at seminars, the most confusing and frustrating concern was the incorporation of test boring logs into the contract document. Endless debate ensued regarding whether these logs should be incorporated into the documents or left in the architect's office for review only by the bidders. The owner who has secured subsurface data via test borings and does not permit the bidders to review that data, can be held liable for fraudulently withholding data which is rightfully due the contractor. If the data is not complete or is erroneous, then the parties responsible for presenting this information can be held negligent. Either way, the parties will find themselves embroiled in litigation which is both time consuming and costly and certainly not desirous on the part of either party.

Test borings are originally taken for design purposes. The structural engineer must know the capacity of the soil prior to designing the foundation and superstructure of the building. Typically, this subsurface data is used by the design professional to inform the bidders regarding subsurface conditions. Although this information is better than nothing, it is certainly not sufficient for a contractor in planning and performing the contractor's work. Test borings for design purposes are taken at strategic locations to give the engineer some idea of the structure required to support the building. The contractor, on the other hand, requires data for the entire site, since the contractor will excavate the area which will be covered by the building and not just the four corners where the test borings may have been taken.

A simple solution to this concern was offered by the late Bob Vansant of Black & Veatch. Bob was instrumental in convincing the owners to spend an additional sum of money to have a construction geo-technical evaluation report prepared along with geo-technical construction documents to provide the contractors with reliable subsurface information. This report, along with the documents, would be incorporated into the bid documents for all bidders to include as part of their bid requirements. Like anything else that is new, this idea was difficult to introduce into the market. However, Bob said that once the geo-technical engineer understood what he was trying to achieve, he liked the idea. Bob also indicated that this additional work added approximately 10 percent to the geo-technical fee.

Contracts may have a "differing site conditions" clause. This clause entitles the contractor to additional compensation and/or time should subsurface work require more time than originally included in the bid. An essential element of this clause is the requirement for timely notice of additional work on the part of the contractor to the owner.

Method and Sequence Changes

§ 6.22 Contractors worth their salt plan their methods of operation and the sequence of activities long before arriving on the site. Steps are then taken to secure the materials, equipment, and labor force necessary to carry out these methods in the sequential order established. If the other party to the contract or a third party is responsible for preventing the evolution of the contractor's plan, then that party can be held liable for the damages incurred by the contractor.

Many contractors have been held liable for damages to the owner if they did not perform the requirements of the contract, even if those requirements were "short of absolute impossibility." The courts have since changed their thinking and have divided "impossibility of performance" into two categories. Actual impossibility is established when the work cannot be performed by any contractor. Practical impossibility is when the work can be performed only under extreme difficulty, expense, injury, or loss. Three conditions must be present in order for impossibility to be applied:

1. The impossibility must be objective. It must inhere in the nature of the act to be performed rather than be personal to the contractor.
2. The facts which make performance impossible must not have been forseeable.
3. The person seeking to be excused from performance must in no way have been responsible for the impossibility.

In the case of *Savage v. Peter Kiewit Sons' Co.*, 432 P.2d 519 (Or. 1967), damages for commercially impractical-to-accomplish actions were denied because the injunction preventing the contractor from continuing his work was forseeable.

The contractor's methods and sequence of work may also be interrupted by interference from either the owner or the architect. When it can be proven that this interference was intentional and not merely negligent, the contractor can sue for damages incurred as a result of the interference.

Delay, Acceleration, and Time Claims

§ 6.23 Claims resulting from delays, constructive acceleration, and other time-related concerns are a major area of litigation in the construction arena. Damages from these claims fall under the category titled "economic loss." These damages differ from other damages because they represent costs incurred by the contractor other than those through property damage or personal injury. Examples of economic losses are lost profits or delay damages. These damages are made up of three types:

1. Actual damages are real, just, and the exact amount of loss.
2. Compensatory damages represent the loss caused by the wrong.
3. Consequential damages represent the indirect loss caused by the wrong.

In some jurisdictions, a third party can be sued for damages resulting from property loss or personal injury, but not for damages for economic loss.

Courts will not permit owners to enjoy the benefits of unjust enrichment. This is a situation where the owner acquires more than what was originally in the contract, due to some error or misunderstanding, but does not want to pay for the additional improvement. If the owner maintains the added work and enjoys the benefit of it, then the owner must pay for it.

In assessing damages against a contractor or subcontractor on behalf of the owner, the courts will employ the "patent-latent" test. This test provides the courts with an opportunity to see who had control over the cause of the damage. If the damage was due to a patent error (one that was obvious to the eye upon normal inspection) then the court will hold the owner responsible. If the cause of the damage is latent (hidden from observation during routine inspection), then the damages will be assessed against the contractor.

Subcontractors' Concerns

§ 6.24 Contractors suffer from losses due to inappropriate methods of payment to their subcontractors and material suppliers, insolvency of subcontractors, and misapplication of funds. In the case of *Weyerhaeuser Co. v. Twin City Millwork Co.*, 191 N.W.2d 40 (Minn. 1971), the contractor paid the subcontractor for 1,000 doors produced and delivered by Weyerhaeuser. Before the sub was able to pay Weyerhaeuser, it declared bankruptcy and Weyerhaeuser sued the prime contractor for payment against the payment bond. The court ruled in favor of Weyerhaeuser and the prime contractor had to pay twice for the doors. In the case of *Western Ready Mix Concrete Co. v. Rodriguez*, 567 P.2d 1118 (Utah 1977), a subcontractor paid a sub for work performed on a bonded project. The subcontractor owed the sub money from a previous project. There was no indication on the check as to which project this payment applied, so the sub's sub assumed it as payment for a previous debt. The sub's sub then sued the prime contractor and the surety for payment. This situation could have been resolved by simply noting on the check the specific job the check covered.

A major concern on a construction site is safety. Although safety affects everyone on the project, including architects, engineers, construction managers, and contractors, there are some unique concerns of the subcontractor regarding safety. Typically, OSHA will hold the organization in charge of the construction activity as the one responsible for maintaining a safe work site. This role is usually performed by the general contractor. With the advent of construction managers, the role can be shifted to the construction manager, based on the written contract. Is the responsibility of safety ever assigned to a subcontractor? For the most part, the responsibility goes to the one who is in control, and seldom is the subcontractor the one in control. However, OSHA has ruled in certain circumstances that if a subcontractor knowingly sends workers into a hazardous and unsafe condition, then that subcontractor is responsible.

To further expand the concern about an injury to a subcontractor's employee, let us consider workers' compensation insurance laws. As we have discussed previously, workers' compensation insurance requirements are a statutory demand made on businesses over a certain size. In return for management's payment of the workers' compensation insurance premiums, it is protected from suit by an injured employee. Now comes a dilemma. A subcontractor knowingly sends a worker into an unsafe condition and the employee is injured. OSHA cites the subcontractor for the unsafe condition. Workers' compensation covers the injured party's damages to a degree. The injured party seeking additional damages files suit against anyone and everyone related to the unsafe condition. Some of those sued may include the architect/engineer, the owner, the prime contractor, the construction manager, and the other subcontractors. The one whom OSHA cited as being the one really responsible is the employer of the injured party, but the injured party cannot sue the subcontractor because of workers' compensation laws. This now leads one to question the validity of such laws. Do these laws promote negligence on the part of the employer? Is the employer of an injured employee unaccountable because of the workers' compensation laws?

However, there are some ways by which this employer can be sued despite the requirements of the worker's compensation laws. One of the other parties sued by the injured party and assessed damages to compensate the injured party may recover losses by suing the injured party's employer in a third party suit. It is at this point where accurate and consistent documentation will benefit the party seeking to recover losses. Some of the major means of documenting an accident are as follows:

1. Photograph the accident or at least the area in which the accident occurred. The photographs should be taken from both a broad view and a very detailed view. When taking detailed photos, relate the detail to a reference landmark so that the viewer will know that this detail is a part of the overall area where the accident took place and not some remote detail that did not affect the accident.
2. Interview eyewitnesses to the accident. Record their observations and candid comments. Be sure to take their names and phone numbers so that they can be reached for further assistance, especially as witnesses on the stand.
3. Record the particular activity which was being performed at the time of the accident and the ramifications stemming from that activity. Include with these records the specific location of the accident, the number of people employed in the area at the time of the accident, the equipment employed and the type of operation, the cause of the accident, and the action taken during and after the accident by those in the immediate area, especially those in charge of the work.
4. Document and record intrinsic conditions including weather conditions and temperature, the time of the day, the date, and visibility.

5. Employ professional consultants to validate the documentation. These consultants can include the photographer, the testing lab technicians, and legal counsel.

If this accurate documentation is properly recorded on file, it can be presented as part of the contractor's testimony in court. In some jurisdictions, the court will assess damages against several parties found to be the cause of the negligence to the degree that each is a part of the cause. In simple language, if three parties are the cause of the negligence, the court will determine what percentage of negligence is attributed to each and the damages will be assessed against each according to that percentage. However, in some jurisdictions, this "contribution" is not recognized and the entire amount of damages is assessed against the party who has the greatest percentage of negligence. Not only will documentation of the accident benefit a party's cause, it will also help determine the degree of negligence, if any, that the court will assess. The more consistent and accurate the documentation, the better the party's position in court.

Final Completion

§ 6.25 Upon completion of the project, the owner requires that the contractor submit certain documents to establish that the work is complete and that the owner can readily take over the building and operate and maintain it accordingly. In addition to maintenance and operation manuals, record drawings, and occupancy permits, the owner also requires the contractor to submit a "release of lien" and "consent of surety." These two documents are crucial to the issuance of final payment to the contractor. As we have discussed previously, a lien is a charge against property for work, labor, services, and material which have increased the value or otherwise benefitted the property. The owner, not wishing to have someone file against the property, requires that the contractor submit a release or waiver of lien for not only the contractor, but also all subcontractors and material vendors. Once this waiver is issued, the contractor and the subs have no claim to lien upon the building.

The consent of surety assures the owner that the project is complete and all the work is acceptable up to the amount charged on the final requisition for payment. If the consent of the surety is not obtained, then the owner runs the risk of having the surety refuse to honor the bond should the contractor default.

Now that the work is complete, the contractors, the subcontractors, and all other organizations involved in the construction process are finished and free to move on to other work. Each state has determined by statute the period of time after the completion of the project that the parties involved will continue their role of responsibility. Most contracts contain a one-year guarantee. Although the statute of limitations will dictate a specific period of time (in some states such as Iowa there is no statute), that does not necessarily mean that the period stipulated is fixed and unchangeable.

One of the first tests that a statute of limitations will encounter is a question of constitutionality. In a Florida case, *Overland Construction Co. v. Sirmons*, 369 So. 2d 572 (Fla. 1979), the state's supreme court found the statute of limitations to be unconstitutional. The court stated that while the statute protected the designers and builders from suits after 12 years, the statute abolished Sirmons' right to sue and provided no alternative form of redress. In simple language, what the court said is that the statute was unfair because it gave the builders and designers an out, but it gave the injured party no other means of compensation.

The next issue concerning statutes of limitations is determining the date of commencement of the time period and therefore the subsequent expiration date. Different courts have determined commencement dates in the following manner:

1. the date on which the negligent act was performed;
2. the date on which the negligent act was discovered; or
3. the date on which the injury occurred.

Therefore, it is almost impossible to determine beforehand how long a claim will be covered by a statute of limitations.

In discussing statutes of limitations, we have not given you any real reason for concern regarding proper documentation. We could say, in a general fashion, that all documentation on the project could be used in a statute of limitations case when the court tries to determine the cause of the damage and the reason for the suit. But strictly speaking, there is no documentation that we can recommend regarding defending against a dismissal under a statute of limitations. There is a case from Kansas which highlights the concerns regarding statutes of limitations. In *Tamarack Development Co. v. Delamater, Freund & Assoc., P.A.*, 675 P.2d 361 (Kan. 1984), the architect Delamater orally agreed to ensure the accuracy of the grade to the developer Tamarack. Two and one half years after the project was complete, Tamarack brought suit against Delamater for breach of contract. The contract that Tamarack referred to was an oral contract whereby the architect guaranteed the accuracy of the grade. The court had to determine whether the architect had breached the oral contract or committed an act of negligence.

The reason for this concern was that the statute of limitations for an action for negligence was two years, whereas the statute of limitations for a breach of contract was a longer period of time. Had the court found against the architect for negligence, then it would have had to dismiss the case on the basis of exceeding the statute of limitations. In its deliberation, the court found against the architect on both counts. Therefore the statute of limitations of two years for a tort action could not be used as a defense by the architect.

A case which brings into focus the statute of limitations, lien laws, provisions in an act which take precedence over a written contract and over a statute (further limiting the statute of limitations), what to document, when to document, and why is *United States v. E.J.T. Construction Co., Inc.* (Del. June 25, 1976—LBAEC 1-10-77). The federal trial court dismissed a subcontractor's lawsuit against the general contractor on the ground that it was barred by a one-year statute of limitations provided for in the Miller Act.

The Miller Act is a law which requires contractors on federal projects to obtain a surety bond to guarantee payments to their subcontractors and suppliers because a contractor cannot lien against federal property. The Miller Act, therefore, presents the parties working on a federal project another avenue for suit against the prime contractor. The subcontractor was not paid and filed a lawsuit to recover payment against the Miller Act bond on September 28, 1973. Calculating the one-year period required by the Miller Act bond, the subcontractor had to prove that labor or materials were supplied to the project after September 28, 1972. The subcontractor was able to accomplish this by introducing evidence that a sub-subcontractor had performed work on the project after that date and that the work was monitored by the subcontractor. Although the courts allowed that the work done by a sub-subcontractor could be used in computing the time period, the court found that the sub-subcontractor, in this case, was correcting deficiencies of work previously done and not performing new work. Since the work was not done in performance of the original contract, the court said it could not be used to compute the statute of limitations time period. As a result, the court dismissed the lawsuit.

Taking a closer look at this case, we can see some of the major factors of concern that a contractor, a subcontractor, and other individuals involved in the construction process, especially on federal work, need to address. First, no matter what the statute of limitations might be for a particular state, if the work is performed for the federal government in that state, or even possibly subsidized by federal government funds, then the statute of limitations is pre-empted by the Miller Act requirements. The next factor is that one must produce evidence, in the form of documentation, that either company personnel or independent subcontractors performed work on a date within the applicable period. The next factor is that liens are not available to those working on federal projects. This includes not only contractors and subcontractors, but any individual. The last factor is whether there was documentation to prove that a sub-subcontractor had worked on a particular date within the period. This documentation must have been in the form of a daily field report which indicated that a sub-subcontractor worked on that date and was supervised by the subcontractor's personnel. This documentation was needed as evidence to secure the subcontractor's position before the court. As we have seen, one must have documentation in order to present his or her position on a legitimate claim before the bench.

APPENDIX 6A

MASTERFORMAT Broadscope Section Titles

(Source: Construction Specifications Institute and Construction Specifications Canada. Reprinted with permission.)

Bidding Requirements, Contract Forms, and Conditions of the Contract

00010	PRE-BID INFORMATION
00100	INSTRUCTIONS TO BIDDERS
00200	INFORMATION AVAILABLE TO BIDDERS
00300	BID FORMS
00400	SUPPLEMENTS TO BID FORMS
00500	AGREEMENT FORMS
00600	BONDS AND CERTIFICATES
00700	GENERAL CONDITIONS
00800	SUPPLEMENTARY CONDITIONS
00900	ADDENDA

Note: The items listed above are not specification sections and are referred to as "Documents" rather than "Sections" in the Master List of Section Titles, Numbers, and Broadscope Section Explanations.

Specifications

DIVISION 1—GENERAL REQUIREMENTS

01010	SUMMARY OF WORK
01020	ALLOWANCES
01025	MEASUREMENT AND PAYMENT
01030	ALTERNATES/ALTERNATIVES
01035	MODIFICATION PROCEDURES
01040	COORDINATION
01050	FIELD ENGINEERING
01060	REGULATORY REQUIREMENTS
01070	IDENTIFICATION SYSTEMS
01090	REFERENCES
01100	SPECIAL PROJECT PROCEDURES
01200	PROJECT MEETINGS
01300	SUBMITTALS
01400	QUALITY CONTROL
01500	CONSTRUCTION FACILITIES AND TEMPORARY CONTROLS
01600	MATERIAL AND EQUIPMENT
01650	FACILITY STARTUP/COMMISSIONING
01700	CONTRACT CLOSEOUT
01800	MAINTENANCE

DIVISION 2—SITEWORK

02010	SUBSURFACE INVESTIGATION
02050	DEMOLITION
02100	SITE PREPARATION

02140	DEWATERING
02150	SHORING AND UNDERPINNING
02160	EXCAVATION SUPPORT SYSTEMS
02170	COFFERDAMS
02200	EARTHWORK
02300	TUNNELING
02350	PILES AND CAISSONS
02450	RAILROAD WORK
02480	MARINE WORK
02500	PAVING AND SURFACING
02600	UTILITY PIPING MATERIALS
02660	WATER DISTRIBUTION
02680	FUEL AND STEAM DISTRIBUTION
02700	SEWERAGE AND DRAINAGE
02760	RESTORATION OF UNDERGROUND PIPE
02770	PONDS AND RESERVOIRS
02780	POWER AND COMMUNICATIONS
02800	SITE IMPROVEMENTS
02900	LANDSCAPING

DIVISION 3—CONCRETE

03100	CONCRETE FORMWORK
03200	CONCRETE REINFORCEMENT
03250	CONCRETE ACCESSORIES
03300	CAST-IN-PLACE CONCRETE
03370	CONCRETE CURING
03400	PRECAST CONCRETE
03500	CEMENTITIOUS DECKS AND TOPPINGS
03600	GROUT
03700	CONCRETE RESTORATION AND CLEANING
03800	MASS CONCRETE

DIVISION 4—MASONRY

04100	MORTAR AND MASONRY GROUT
04150	MASONRY ACCESSORIES
04200	UNIT MASONRY
04400	STONE
04500	MASONRY RESTORATION AND CLEANING
04550	REFRACTORIES
04600	CORROSION RESISTANT MASONRY
04700	SIMULATED MASONRY

DIVISION 5—METALS

05010	METAL MATERIALS
05030	METAL COATINGS
05050	METAL FASTENING
05100	STRUCTURAL METAL FRAMING
05200	METAL JOISTS
05300	METAL DECKING
05400	COLD FORMED METAL FRAMING
05500	METAL FABRICATIONS
05580	SHEET METAL FABRICATIONS

05700	ORNAMENTAL METAL
05800	EXPANSION CONTROL
05900	HYDRAULIC STRUCTURES

DIVISION 6—WOOD AND PLASTICS

06050	FASTENERS AND ADHESIVES
06100	ROUGH CARPENTRY
06130	HEAVY TIMBER CONSTRUCTION
06150	WOOD AND METAL SYSTEMS
06170	PREFABRICATED STRUCTURAL WOOD
06200	FINISH CARPENTRY
06300	WOOD TREATMENT
06400	ARCHITECTURAL WOODWORK
06500	STRUCTURAL PLASTICS
06600	PLASTIC FABRICATIONS
06650	SOLID POLYMER FABRICATIONS

DIVISION 7—THERMAL AND MOISTURE PROTECTION

07100	WATERPROOFING
07150	DAMPPROOFING
07180	WATER REPELLENTS
07190	VAPOR RETARDERS
07195	AIR BARRIERS
07200	INSULATION
07240	EXTERIOR INSULATION AND FINISH SYSTEMS
07250	FIREPROOFING
07270	FIRESTOPPING
07300	SHINGLES AND ROOFING TILES
07400	MANUFACTURED ROOFING AND SIDING
07480	EXTERIOR WALL ASSEMBLIES
07500	MEMBRANE ROOFING
07570	TRAFFIC COATINGS
07600	FLASHING AND SHEET METAL
07700	ROOF SPECIALTIES AND ACCESSORIES
07800	SKYLIGHTS
07900	JOINT SEALERS

DIVISION 8—DOORS AND WINDOWS

08100	METAL DOORS AND FRAMES
08200	WOOD AND PLASTIC DOORS
08250	DOOR OPENING ASSEMBLIES
08300	SPECIAL DOORS
08400	ENTRANCES AND STOREFRONTS
08500	METAL WINDOWS
08600	WOOD AND PLASTIC WINDOWS
08650	SPECIAL WINDOWS
08700	HARDWARE
08800	GLAZING
08900	GLAZED CURTAIN WALLS

DIVISION 9—FINISHES

09100	METAL SUPPORT SYSTEMS
09200	LATH AND PLASTER

09250 GYPSUM BOARD
09300 TILE
09400 TERRAZZO
09450 STONE FACING
09500 ACOUSTICAL TREATMENT
09540 SPECIAL WALL SURFACES
09545 SPECIAL CEILING SURFACES
09550 WOOD FLOORING
09600 STONE FLOORING
09630 UNIT MASONRY FLOORING
09650 RESILIENT FLOORING
09680 CARPET
09700 SPECIAL FLOORING
09780 FLOOR TREATMENT
09800 SPECIAL COATINGS
09900 PAINTING
09950 WALL COVERINGS

DIVISION 10—SPECIALTIES

10100 VISUAL DISPLAY BOARDS
10150 COMPARTMENTS AND CUBICLES
10200 LOUVERS AND VENTS
10240 GRILLES AND SCREENS
10250 SERVICE WALL SYSTEMS
10260 WALL AND CORNER GUARDS
10270 ACCESS FLOORING
10290 PEST CONTROL
10300 FIREPLACES AND STOVES
10340 MANUFACTURED EXTERIOR SPECIALTIES
10350 FLAGPOLES
10400 IDENTIFYING DEVICES
10450 PEDESTRIAN CONTROL DEVICES
10500 LOCKERS
10520 FIRE PROTECTION SPECIALTIES
10530 PROTECTIVE COVERS
10550 POSTAL SPECIALTIES
10600 PARTITIONS
10650 OPERABLE PARTITIONS
10670 STORAGE SHELVING
10700 EXTERIOR PROTECTION DEVICES FOR OPENINGS
10750 TELEPHONE SPECIALTIES
10800 TOILET AND BATH ACCESSORIES
10880 SCALES
10900 WARDROBE AND CLOSET SPECIALTIES

DIVISION 11—EQUIPMENT

11010 MAINTENANCE EQUIPMENT
11020 SECURITY AND VAULT EQUIPMENT
11030 TELLER AND SERVICE EQUIPMENT
11040 ECCLESIASTICAL EQUIPMENT
11050 LIBRARY EQUIPMENT
11060 THEATER AND STAGE EQUIPMENT

11070 INSTRUMENTAL EQUIPMENT
11080 REGISTRATION EQUIPMENT
11090 CHECKROOM EQUIPMENT
11100 MERCANTILE EQUIPMENT
11110 COMMERCIAL LAUNDRY AND DRY CLEANING EQUIPMENT
11120 VENDING EQUIPMENT
11130 AUDIO-VISUAL EQUIPMENT
11140 VEHICLE SERVICE EQUIPMENT
11150 PARKING CONTROL EQUIPMENT
11160 LOADING DOCK EQUIPMENT
11170 SOLID WASTE HANDLING EQUIPMENT
11190 DETENTION EQUIPMENT
11200 WATER SUPPLY AND TREATMENT EQUIPMENT
11280 HYDRAULIC GATES AND VALVES
11300 FLUID WASTE TREATMENT AND DISPOSAL EQUIPMENT
11400 FOOD SERVICE EQUIPMENT
11450 RESIDENTIAL EQUIPMENT
11460 UNIT KITCHENS
11470 DARKROOM EQUIPMENT
11480 ATHLETIC, RECREATIONAL, AND THERAPEUTIC EQUIPMENT
11500 INDUSTRIAL AND PROCESS EQUIPMENT
11600 LABORATORY EQUIPMENT
11650 PLANETARIUM EQUIPMENT
11660 OBSERVATORY EQUIPMENT
11680 OFFICE EQUIPMENT
11700 MEDICAL EQUIPMENT
11780 MORTUARY EQUIPMENT
11850 NAVIGATION EQUIPMENT
11870 AGRICULTURAL EQUIPMENT

DIVISION 12—FURNISHINGS

12050 FABRICS
12100 ARTWORK
12300 MANUFACTURED CASEWORK
12500 WINDOW TREATMENT
12600 FURNITURE AND ACCESSORIES
12670 RUGS AND MATS
12700 MULTIPLE SEATING
12800 INTERIOR PLANTS AND PLANTERS

DIVISION 13—SPECIAL CONSTRUCTION

13010 AIR SUPPORTED STRUCTURES
13020 INTEGRATED ASSEMBLIES
13030 SPECIAL PURPOSE ROOMS
13080 SOUND, VIBRATION, AND SEISMIC CONTROL
13090 RADIATION PROTECTION
13100 NUCLEAR REACTORS
13120 PRE-ENGINEERED STRUCTURES
13150 AQUATIC FACILITIES
13175 ICE RINKS
13180 SITE CONSTRUCTED INCINERATORS
13185 KENNELS AND ANIMAL SHELTERS

13200 LIQUID AND GAS STORAGE TANKS
13220 FILTER UNDERDRAINS AND MEDIA
13230 DIGESTER COVERS AND APPURTENANCES
13240 OXYGENATION SYSTEMS
13260 SLUDGE CONDITIONING SYSTEMS
13300 UTILITY CONTROL SYSTEMS
13400 INDUSTRIAL AND PROCESS CONTROL SYSTEMS
13500 RECORDING INSTRUMENTATION
13550 TRANSPORTATION CONTROL INSTRUMENTATION
13600 SOLAR ENERGY SYSTEMS
13700 WIND ENERGY SYSTEMS
13750 COGENERATION SYSTEMS
13800 BUILDING AUTOMATION SYSTEMS
13900 FIRE SUPPRESSION AND SUPERVISORY SYSTEMS
13950 SPECIAL SECURITY CONSTRUCTION

DIVISION 14—CONVEYING SYSTEMS

14100 DUMBWAITERS
14200 ELEVATORS
14300 ESCALATORS AND MOVING WALKS
14400 LIFTS
14500 MATERIAL HANDLING SYSTEMS
14600 HOISTS AND CRANES
14700 TURNTABLES
14800 SCAFFOLDING
14900 TRANSPORTATION SYSTEMS

DIVISION 15—MECHANICAL

15050 BASIC MECHANICAL MATERIALS AND METHODS
15250 MECHANICAL INSULATION
15300 FIRE PROTECTION
15400 PLUMBING
15500 HEATING, VENTILATING, AND AIR CONDITIONING (HVAC)
15550 HEAT GENERATION
15650 REFRIGERATION
15750 HEAT TRANSFER
15850 AIR HANDLING
15880 AIR DISTRIBUTION
15950 CONTROLS
15990 TESTING, ADJUSTING, AND BALANCING

DIVISION 16—ELECTRICAL

16050 BASIC ELECTRICAL MATERIALS AND METHODS
16200 POWER GENERATION—BUILT-UP SYSTEMS
16300 MEDIUM VOLTAGE DISTRIBUTION
16400 SERVICE AND DISTRIBUTION
16500 LIGHTING
16600 SPECIAL SYSTEMS
16700 COMMUNICATIONS
16850 ELECTRIC RESISTANCE HEATING
16900 CONTROLS
16950 TESTING

CHAPTER 7

Documentation for Quality Assurance and Quality Control

QUALITY ASSURANCE PROGRAM DEVELOPMENT

§ 7.1 In the nuclear industry, documentation of construction quality control has been developed to the most exacting standards. An important precedent for the development of a quality assurance program is found in the ANSI/ASME NQA-1 standard, "Quality Assurance Program Requirements for Nuclear Facilities," which encompasses the following elements:

- qualification of inspection, examination, and testing personnel;
- requirements for the collection and maintenance of quality assurance records;
- quality assurance terms and definitions;
- auditing of quality assurance programs; and
- quality assurance for control of procurement.

QUALITY ASSURANCE PLAN FOR CONSTRUCTION MANAGER

§ 7.2 Appendix 7A at the end of this chapter is the generic quality assurance plan for construction management projects implemented by O'Brien-Kreitzberg & Associates, Inc. This plan must be customized for each assignment. In most cases, this customization deletes the scope section because the owner does not desire that level of professional services or is providing that scope in some other fashion.

QUALITY CONTROL PROGRAM

§ 7.3 The quality control program usually includes field testing. Traditionally, test requirements are described in the specifications wherever (and whenever) the design engineer (or architect) chooses. The quality assurance procedures should call for an index of all test requirements (or the requirements themselves) to be listed in Division 1 of the specifications under § 01400, "Quality Control."

A listing of the mediumscope section titles in Division 1, "General Requirements," of the Construction Specifications Institute's MASTERFORMAT can be utilized to describe all, or part, of the supplemental conditions. It can also be used as a checklist for quality assurance to review the draft general conditions/supplemental conditions to ensure inclusion of all necessary elements.

If an index of all test requirements has not been prepared during the design phase, the quality control group must set up an inspection plan. Step one in the inspection plan is the specification review, and the writing of each test requirement. Figure 7-1 shows a typical form that can be used to record individual tests. The individual requirements should then be summarized by area or floor, similar to Figure 7-2. The contract may require the contractor to conduct the quality control tests with observation by the inspection team, or the inspection team may take samples and conduct tests.

To accomplish job site testing, appropriate reference documents must be available, since the specifications usually refer to testing procedures by reference to standards of the American Society for Testing and Materials, American Concrete Institute, American Society of Mechanical Engineers, American National Standards Institute, or others. The inspection team should review specifications and be certain that referenced standards are available and have been reviewed before testing is required. It is often appropriate to conduct a trial or rehearsal test before actual job materials are mixed.

Often testing may be done under the auspices of the materials manufacturer, although the obvious vested interest involved must be considered. Architect-engineers often recommend the use of special testing laboratories or groups as an adjunct to the inspection team. This practice can be both effective and economical, as it limits the amount of specialized testing equipment that must be purchased and does not require

EQUIPMENT FIELD TEST REQUIREMENTS

O'BRIEN–KREITZBERG & ASSOC. INC.
CAMDEN COUNTY MUNICIPAL UTILITIES AUTHORITY
DELAWARE NO. 1 WPCF IMPROVEMENTS
CONTRACT 160

SPEC. SECT:______________________ SUB. SECT: ______________________

STRUCTURE: ______________________

LOCATION: ______________________

MANUFACTURER: ______________________

EQUIPMENT: ______________________

EQUIP. NO: ______________________

MANUFACTURERS REP. REQUIRED

☐ YES ☐ NO NAME: ______________________

ACTION	RESULT	ACCEPTABLE	
		YES	NO

REMARKS

____

INSPECTOR ______________________

Figure 7–1

CCMUA — DELAWARE NO. 1 START UP
STRUCTURE LEAKAGE TESTING

	UNIT											
	Aeration Tank No. 5	Hydraulic	Ready Test	Tested		Air						
	Aeration Tank No. 6	Hydraulic				Air						
	Aeration Tank No. 7	Hydraulic				Air						
	Aeration Tank No. 8	Hydraulic				Air						
	Aeration Tank New Effluent Channel											
	Cross Channel No. 3											
	Cross channel No. 4											
	Final Sedimentation Tank Influent Channel											
	Final Sedimentation Tank No. 5											
	Final Sedimentation Tank No. 6											
	Final Sedimentation Tank No. 7											
	Final Sedimentation Tank No. 8											
	Chlorine Contact Tank Channels											
	Drain Sump — FST Operating Gallery											
	East Scum Well — FST Pipe Gallery											
	West Scum Well — FST Pipe Gallery											

Figure 7-2

the inspection team to develop special skills which are used only to a limited extent on a project. Conversely, where large amounts of material are used on the job, the inspection team may learn the skills to conduct specific tests. For instance, on large excavation or backfill projects, certain special moisture content tests are run so frequently that the field team could conduct them.

In cases where materials have been mixed off the job site, such as ready-mix concrete, a special test team may be assigned to the assembly or batching area. Typically, certified laboratory services are retained to perform the tests at the batch plant. This procedure is particularly appropriate for premixed concrete, asphalt, gravel, and aggregate.

Off-site testing also is used for the inspection of materials at the point of manufacture. Usually, this is during quality control testing or as required (for example, water pressure testing). Usually the test is conducted using the manufacturer's equipment and personnel under the supervision of the inspection team.

The test requirements for construction sites are many and varied. The following sections will describe some typical procedures and equipment, but the listing is not inclusive. The specifications will provide explicit directions to the inspection team for the type of test to be run, equipment to be used, and conditions of the test.

Soil Testing

§ 7.4 Native soil may be checked for loadbearing capacity through actual static loading or by taking under-turf samples for tests. Backfill samples are checked to determine the degree of compaction, which in turn is a measure of the loadbearing capacity of backfill soil.

The Anteus consolidometer is an accurate instrument designed for testing of soil consolidation. The unit can also be adapted to test porosity, measuring the ability of the soil to resist hydrostatic pressure. One of the most recent devices for measuring proctor values for soil and soil aggregate uses a nuclear source and detectors for the nondestructive determination of soil values. This is done by a family of instruments made by Troxler and known as the 3400D series of surface moisture-density gauges. The devices include solid-state semiconductor components and liquid crystal display readouts. The units are battery powered and contain a micro-computer which holds the calibration constants and algorithms necessary to compute and display wet density, moisture, dry density, percent moisture, and percent compaction.

When these instruments are used to measure soil characteristics, a direct transmission mode can be used that places the gamma source into the material by means of a punched access hole. Standard gauges have an 8-inch depth capability in 2-inch increments. (Other increments and depths are available.) American Society for Testing Materials Standard D-2922-78, "Standard Test Method for Density of Soil and Soil Aggregate in Place by Nuclear Methods," applies for in-place soil tests.

Documentation

§ 7.5 Testing is a time-consuming, serious, expensive, and necessary operation. If test results are not recorded appropriately, retesting may be required—and often may be impossible. Figure 7-3 is a typical form used to record piping tests related to the service pressures of process materials. Figure 7-4 is a partially executed version of the test plan. During testing, deficiencies will be discovered. Figure 7-5 is a deficiency report, and Figure 7-6 is a deficiency summary report. These are important forms for the recording of rehabilitation/remedial work. Figure 7-7 is a form prepared by the resident inspector for all field tests of water, sewer, and steam lines, as well as other major equipment installations. Appendix 7B sets forth some of the forms used as a general conditions appendix to the New York Transit Authority quality assurance procedures.

Figure 7–3

INSPECTOR'S TEST REPORT

Date ______________________ Project ______________________
Contract No. ______________________ Proj. area ______________________
Witness to test: Signature:
Contractor ______________________ ______________________
Inspector ______________________ ______________________
Type of test (press., leakage, etc.) ______________________
System, line or structure identification:
Number or title: ______________________
Location (co-ordinates, etc.): ______________________
Reference drawing: ______________________
PRESSURE TEST (PIPING)
BURIED CAST-IRON (ANSI/AWWA C-600 Section 4)
Test media: ______________________
Pressure specified: ______________ PSI
Pressure actual – start: ______________ PSI finish: ______________ PSI
Allowable leakage: (calculated)

$$L = \frac{SD \quad P}{133200}$$

Where — L = Allowable Leakage (gal/hr)
S = Length of Pipe (ft)
D = Nominal Diameter (in)
P = Average Test Pressure (PSIG)

CALCULATION
Pipe length __________ Pipe Dia. __________ Allowable Leakage __________
Pipe length __________ Pipe Dia. __________ Allowable Leakage __________
Pipe length __________ Pipe Dia. __________ Allowable Leakage __________
Total actual leakage ______________
Duration: Time started: ______________ Time finished: ______________
Elapsed time: ______________
Comments ______________________
____
____
(Ductile Iron, Gray Cast Iron, Mechanical Jointed, and Push on Jointed pipe included in this section.)
SMOKE TEST (CCMUA W & M 26)
Test subject: (structure, pipe, etc.) ______________________
Bombs used: (type, number, etc.) ______________________
Observations: ______________________
____

Comments____

____

PRESSURE TEST (Piping – all except buried cast iron)

Test media: ____

Pressure specified:________________PSI actual: ________________PSI

Allowable leakage:______________________________per applicable specs

Duration: Time started: ________________ Time finished: ____________

Time elapsed: Specified: __________ Actual:________________

Line elevation: High point:________________ At gauge:________________

Comments____

____

LEAKAGE TEST

Structure/vessel/etc.: ________________________________

Test media: ____

Normal Operating level: ______________________________

Test level: Start:______________________ Level:________________

*Allowable loss: Specified:________________ Actual:________________

Duration: Time started: ________________ Time finished: ____________

Time elapsed: Specified: __________ Actual:________________

Comments____

____

*CCMUA W & M 26

INFILTRATION OR LEAKAGE TEST (Drains & sewers)

Test media: ____

Line elevation: High point:________________ Low point: ______________

Duration: Time started: ________________ Time finished: ____________

Time elapsed: Specified: __________ Actual:________________

Gauge or level readings: Start: ______________ 30 minutes: ____________

60 minutes: __________ 90 minutes: ____________

120 minutes: __________ 150 minutes: ____________

180 minutes: __________ 210 minutes: ____________

240 minutes: __________ (min. time limit)

Allowable loss: (CCMUA W & M 26.04) ________________________

Actual loss: ____

Comments____

____

Test approved:____________________ Test disapproved: ________________

Reason:______________________

	UNIT											
	Aeration Tank No. 5	Hydraulic	Ready Test 5-26	Tested 6-2.3		Air	Ready Test 7-1-87	Tested 7-2-87				
	Aeration Tank No. 6	Hydraulic	5-26	6-1.2		Air	7-1-87					
	Aeration Tank No. 7	Hydraulic	5-26	6-1.2		Air	7-1-87	7-3-87				
	Aeration Tank No. 8	Hydraulic	5-26	6-1.2		Air	7-1-87					
	Aeration Tank New Effluent Channel											
	Cross Channel No. 3											
	Cross channel No. 4											
	Final Sedimentation Tank Influent Channel											
	Final Sedimentation Tank No. 5	Tank Filling	6-2	6-5, 6-9								
	Final Sedimentation Tank No. 6	Tank Filling 6-2	6-5	6-5, 6-9								
	Final Sedimentation Tank No. 7		6-15	7-3, 7-6								
	Final Sedimentation Tank No. 8		6-19	—								
	Chlorine Contact Tank Channels	Tank Full										
	Drain Sump — FST Operating Gallery	Sump Full										
	East Scum Well — FST Pipe Gallery											
	West Scum Well — FST Pipe Gallery											

Figure 7-4

No. 0D-160–________

OPERATIONAL DEFICIENCY

CAMDEN COUNTY MUNICIPAL UTILITIES AUTHORITY
DELAWARE NO.1 – WPCF IMPROVEMENTS
CCMUA PROJECT NO. DIII – 2160

LOCATION:______________________________DATE REQUESTED:____________

DESCRIPTION OF DEFICIENCY AND REQUIRED ACTION:

____

____

____

____

____

TYPE OF WORK: CONTRACT ☐ WARRANTY ☐

PRIORITY: ASAP ☐ ROUTINE ☐

REQUESTED BY: ____________
CCMUA

PAUL A. LAURENCE CO.:____________________ DATE CORRECTED: ____________
FRANK X. BEACOM

VERIFIED: ______________________________________DATE: ____________
O'BRIEN – KREITZBERG & ASSOC., INC.

______________________________________DATE:____________
CAMDEN COUNTY MUNICIPAL UTILITIES AUTHORITY

DEFICIENCY CORRECTED: YES ☐ NO ☐

FOLLOW–UP REQUIRED: ____

____

____

____

Figure 7-5

CAMDEN COUNTY MUNICIPAL UTILITIES AUTHORITY

DELAWARE NO. 1 WPCF IMPROVEMENTS, OPERATIONAL DEFICIENCY LOG

PAGE_____OF_____

CONTRACT 160

EPA PROJECT NO. C-34-708-06 CCMUA PROJECT NO. DIII-2160

OPERATIONAL DEFICIENCY NO.	DESCRIPTION	TYPE OF WORK	PRIORITY	DATE CORRECTED	DATE VERIFIED		DEFICIENCY CORRECTED	REMARKS
					OKA	CCMUA		

Figure 7-6

FIELD TEST
IDENTIFICATION

Project_______________

Contract_______________

1. Exact Location of Test on Project ______________________________
2. What is being Tested______________________________________

   ______________________________ Date of Test _____________
3. Paragraph Numbers in Specification and/or reason for making test.

   ____

   ____
4. Exactly how test was made, Instruments, Gauges, etc. used.

   ____

   ____

   ____

   ____
5. Complete list of the personnel present witnessing test and their title.

   ____

   ____
6. Comments: ____

   ____

   ____

Figure 7-7

APPENDIX 7A

O'BRIEN-KREITZBERG & ASSOC., INC.

POLICY STATEMENT

This Quality Assurance Program Plan establishes the O'Brien-Kreitzberg & Associates, Inc. (OKA) Quality Assurance Program, which is intended to control the quality aspects of construction management and construction inspection services provided by OKA. The specific applicability of the requirements of this Program Plan shall be established in contract and project documents.

OKA personnel shall effectively implement this Program Plan, as a primary commitment to assure the quality of construction.

The Principal-in-Charge of each project shall regularly review the status and adequacy of the project quality program. Personnel responsible for project quality activities shall have free access to the Principal-in-Charge to identify quality problems.

James J. O'Brien, P.E. Fred C. Kreitzberg, P.E.

OKA QUALITY ASSURANCE PROGRAM PLAN
TABLE OF CONTENTS

1.0 QUALITY PROGRAM SCOPE

1.1 Construction Management and Inspection Services

1.1.1 This Quality Assurance Program Plan establishes the Quality Assurance Program for construction management and inspection services provided by O'Brien-Kreitzberg & Associates, Inc. (OKA).

1.1.2 The OKA commitment to excellence in construction management includes vigorous controls for cost and schedule, which, although compatible with and supported by the quality requirements in this Program Plan, are established and implemented in separate corporate and project documents.

1.2 Quality Assurance Program Implementation

1.2.1 The specific application of the requirements of this Program Plan for a project shall be defined by contract documents, which shall establish:

- Project organization and assignment of responsibilities.
- Definition of external interfacing quality organizations and functions.
- Project-unique requirements which supplement the Program Plan for a specific project.

1.2.2 The Quality Assurance Program is implemented in corporate and project procedures, specifications and instructions.

1.2.3 Corporate procedures are approved by the Principal-in-Charge. Project procedures and instructions are approved by the Project Manager.

1.2.4 The Project Manager shall provide for preparation, listing, and distribution of applicable project control documents to project personnel.

1.2.5 The Project Manager shall provide for the indoctrination and training of project personnel in the requirements and application of the project control documents.

1.3 Definitions

The word "shall" is used to denote a requirement. The word "should" is used to denote a recommendation. The word "may" is used to denote permission; neither a requirement nor a recommendation.

2.0 APPLICABLE STANDARDS

2.1 The OKA Quality Assurance Program Plan is responsive to the requirements of the following standards, as they apply to the construction management and construction inspection services offered by OKA:

- ANSI/ASME N45.2-1977, "Quality Assurance Program Requirements for Nuclear Facilities."
- Military Specification MIL-Q-9858A of 12/14/63, "Quality Program Requirements."
- Military Specification MIL-I-45208A of 12/16/63, "Inspection System Requirements."
- RDT F2-2/1973, "Quality Assurance Program Requirements."

3.0 QUALITY PROGRAM MANAGEMENT

3.1 **Organization** [3.1]

3.1.1 A typical project organization for construction management and inspection is shown below:

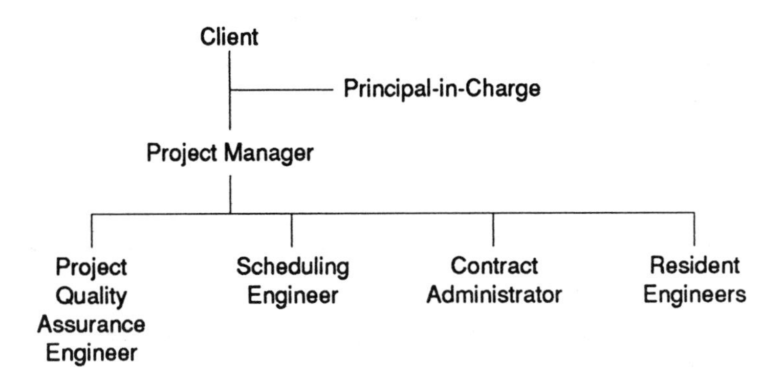

3.1.2 For each project, specific authority and lines of communication are established to report, control and resolve problems that could affect the quality of the work effort. The project organization and a description of assigned responsibilities shall be documented and maintained current in the Project Procedures Manual.

3.2 **Management Review and Audit**

3.2.1 The Principal-in-Charge shall regularly review the adequacy of the project quality program, and implement necessary program additions or changes. This review may be accomplished by:

- Review of project reports.
- Discussions with project and client personnel.
- Formal audit by an independent audit team.

3.2.2 Formal audits shall include an evaluation of quality assurance practices, procedures, and instructions; the effectiveness of implementation; and conformance with policy directives. In performing this evaluation, the audits should include evaluation of work areas, activities, processes, and items; and review of documents and records.

3.2.3 The Principal-in-Charge shall define the scope of the audit and select an audit team experienced in auditing techniques, competent in the technical areas to be evaluated, and independent of project responsibilities. The audit shall be planned and accomplished in accordance with written procedures or checklists, with the results reported to the Principal-in-Charge, who is responsible for initiating and evaluating corrective action.

3.3 **Construction and Inspection Planning**

3.3.1 The Project Manager shall provide for the review of construction drawings and specifications with respect to:

- Constructability, including, but not limited to, conformance to generally accepted construction practices, enforceability of specifications and avoidance of disputes.
- Planning of inspection and testing requirements, methodology and documentation.

3.3.2 Inspection planning shall include:

- Construction/Inspection sequencing and identification of mandatory hold points.
- Inspection procedure, personnel qualification and equipment requirements.
- Scheduling inspector training and certification.
- Identifying inspection and testing to be subcontracted.
- Scheduling inspections and tests.
- Identifying individuals authorized to request and to approve inspections and tests.

3.4 **Inspection Procedures**

3.4.1 The Project Quality Assurance Engineer shall prepare inspection procedures which establish:

- the characteristics to be inspected.
- the inspection methods.
- the acceptance and rejection criteria.
- the methods for recording inspection results.
- special preparation, cleaning, or measuring devices.

3.5 **Inspection Records**

3.5.1 Inspection and testing records shall be identified, collected and indexed to assure retrievability.

3.5.2 The records shall include the results of reviews, inspections, tests, audits, monitoring of work performance, materials analyses, and inspection logs. The records shall also include, as appropriate, closely related data such as qualifications of personnel, procedures, and equipment. Inspection and test records shall, as a minimum, identify the date of inspection or test, the inspector or data recorder, the type of observation, the results, the acceptability, and the action taken in connection with any deficiencies noted. Required records shall be legible, identifiable, and retrievable.

3.5.3 These records shall be reviewed to assure that the records are legible and complete.

3.5.4 The records which have been identified and collected shall be suitably protected against fire, theft, and damage.

3.5.5 The Project Quality Assurance Engineer shall identify, collect, review, index, maintain and arrange for the transfer of inspection records.

3.5.6 Inspection records shall be transferred to the client in accordance with project contract documents. Records may be transferred at various points in the project and at the end of the project. The Project Manager shall obtain the specific consent of the client prior to the destruction of any inspection records.

3.6 Corrective Action

3.6.1 Conditions Adverse to Quality

Conditions adverse to quality may be identified by a number of techniques:

- Audits of OKA by clients.
- Internal OKA audits.
- Audits of sub-tier suppliers by OKA.
- Project reports.
- Principal-in-Charge review of projects.
- Discrepancy reports.

Each of the above techniques has a mechanism to effect the correction of the condition adverse to quality: the audit technique has the audit report and response to the audit report mechanism, the project report and the Principal-in-Charge reviews result in management action; and the discrepancy report has the disposition mechanism.

3.6.2 Significant Conditions Adverse to Quality

Significant conditions adverse to quality are those which extend beyond a single condition or item. A significant condition adverse to quality must be generic in nature to a large number of items or be a deficiency in quality program.

Each condition adverse to quality shall be analyzed to determine if it represents a significant condition adverse to quality, as defined above. This analysis shall be performed by the individual making the disposition of the condition adverse to quality.

The Principal-in-Charge shall perform an analysis to determine if there are any broad programmatic problem areas or if any negative trends are detectable. This analysis shall be performed at lease annually, as part of the management review.

4.0 CONSTRUCTION DOCUMENTS AND STANDARDS

4.1 Construction Documents and Standards

4.1.1 The Project Manager shall establish procedures to control design documents used for construction inspection activities, including drawings, specifications, procedures and instructions, and changes to these documents, to preclude the use of unapproved or out-dated documents. These procedures shall control:

- Defining the issuing authority for various documents and changes.
- Establishing and updating distribution lists.

- Verifying the use of current documents for inspection activities.
- Removing obsolete drawings from use.

4.1.2 The inspection program (discussed in Section 6.3) shall include monitoring design document controls established by construction contractors, to verify that current documents and document changes are issued and used to control construction activities.

4.2 **Design Changes**

4.2.1 The Project Manager shall establish procedures to control engineering change request, approval, and issue for construction, in conformance with project contract requirements. These procedures shall include monitoring implementation of approved engineering changes.

4.2.2 The inspection program (discussed in Section 6.3) shall include monitoring field changes accomplished by construction contractors, for compliance with applicable construction specification requirements. The verification of as-built drawings (discussed in section 4.3) shall include these field changes.

4.3 **As-Built Records**

4.3.1 The Project Quality Assurance Engineer shall provide for the collection and verification of as-built records as required by the construction specifications and contract documents.

4.3.2 Final as-built drawings shall include all approved engineering and field changes.

4.3.3 In addition to as-built drawings, as-built records include specifications, procedures and instructions used in control of configurations or in construction, inspection records (discussed in Section 3.5), and material certifications and test data.

4.3.4 The Project Quality Assurance Engineer shall provide for the indexing and transfer of as-built records to the client upon completion of the project, as required by the construction specifications.

4.4 **Inspection and Test Equipment Calibration**

4.4.1 The Project Quality Assurance Engineer shall provide for the control, calibration and adjustment of inspection and test equipment, to assure that tools, gauges, instruments, and other measuring and testing devices used for inspection and testing are properly controlled, calibrated, and adjusted at specific periods to maintain accuracy within necessary limits. These requirements are not intended to imply a need for special calibration and control measures on rulers, tape measures, levels, and such other devices, if normal commercial practices provide adequate accuracy.

4.4.2 The calibration of inspection and testing equipment shall be accomplished in accordance with written procedures, which shall include the following requirements:

- Identification of equipment and traceability to calibration data.
- Calibration methods, frequency, maintenance, and control.
- Labeling and marking of equipment to indicate due date for next calibration.

- Provisions for determining the validity of previous measurements when equipment is determined to be out of calibration.
- Use of calibration standards with an uncertainty (error) of less than one-fourth the tolerance of equipment being calibrated, within the state-of-the-art.
- Traceability of reference and transfer standards to nationally recognized standards. When national standards do not exist, the basis for calibration shall be documented.

4.4.3 Calibration intervals shall be based on required accuracy, use of equipment, stability characteristics, or other factors affecting the measurement. Calibration may be performed on site or by qualified laboratories utilizing competent personnel. Equipment which is found to be frequently out of adjustment shall be repaired or replaced.

4.4.4 Special calibration shall be performed when the accuracy of the equipment is suspect. When inspection or test equipment is found to be out of calibration, an evaluation shall be made and documented of the validity of previous inspection or test results and of the acceptability of items previously inspected or tested.

4.4.5 These requirements for inspection and test equipment calibration shall be imposed on sub-tier suppliers, including inspection and testing services.

4.4.6 Installed instrumentation used in acceptance testing shall be calibrated in accordance with project acceptance testing documents.

5.0 CONTROL OF PURCHASES

5.1 Supplier Evaluation

5.1.1 The Project Manager shall provide for the evaluation of the capability of a supplier to provide an item or service in accordance with engineering and quality requirements. This evaluation shall be based on one or more of the following:

- Supplier's capability to comply with the elements of the quality standard applicable to the type of material, equipment, or service being procured.
- Past records and performance for similar procurements to ascertain the capability of supplying a manufactured product (or services) under an acceptable quality assurance system.
- Surveys of supplier's facilities and quality assurance program to determine his capability to supply a product which satisfies the design, manufacturing, and quality requirements. This survey should include, as appropriate, facilities, production capabilities, personnel capabilities, process and inspection capabilities, and organization, in addition to the supplier's quality assurance program.

5.2 Control of Sub-Tier Suppliers

5.2.1 The Project Quality Assurance Engineer shall review purchase orders for items and services supporting construction inspection activities, to verify:

- complete and correct statement of the technical and quality requirements, incluing reference to appropriate standards and specifications.

- identification of records to be prepared, maintained, submitted or made available for review, such as drawings, specifications, procedures, procurement documents, inspection and test records, personnel and procedure qualifications, and material, chemical, and physical test results. Record retention and disposition requirements shall be provided.

- provisions for extending applicable requirements of procurement documents to lower tier subcontractors and suppliers, including purchaser's access to facilities and records, if appropriate.

5.2.2 The Project Quality Assurance Engineer shall assess the effectiveness of the control of quality by suppliers supporting construction inspection activities. This assessment shall be accomplished at intervals consistent with the complexity of the item or service, with the quantity of material furnished, and with the duration of the service furnished, based on one or more of the following:

- Direct source inspection (discussed in Section 5.3).

- Reviews of objective evidence of quality furnished by the supplier, such as inspection and test records, personnel and procedure qualifications, material physical and chemical test results, and supplier licensing and certification.

- Periodic audits of the supplier quality program and procedures.

- Comparison or re-test of supplier products or services, by independent testing facilities, and/or against known standards.

5.3 **Source Inspection**

5.3.1 When specified in project contract documents, the Project Manager shall establish procedures for source inspection at supplier's facilities. The source inspection activities may include:

- Reviewing material acceptability, including associated expendable and consumable materials necessary for the functional performance of structures, systems, and components.

- Witnessing in-process inspections, tests, and non-destructive examinations.

- Reviewing the qualification of procedures, equipment, and personnel.

- Verifying that fabrication or construction procedures and processes have been approved and are properly applied.

- Verifying the implementation of the quality assurance/quality control systems.

- Reviewing document packages for compliance to procurement document requirements, including qualifications, process records, inspection and test records.

- Reviewing certificates of compliance for adequacy.

- Verifying that nonconformances have been properly controlled.

5.3.2 The results of source inspection activities shall be documented and shall include copies of certifications, chemical and physical analyses, inspection reports, test results, personnel and process qualification results, code stampings, and nondestructive test reports as required by the applicable specification.

6.0 CONSTRUCTION CONTROL AND INSPECTION

6.1 Material and Inventory Controls

6.1.1 The inspection program (discussed in Section 6.3) shall include monitoring material controls utilized by construction contractors, to verify that these controls are implemented in compliance with construction specification requirements. These material controls include:

- Receiving and receiving inspection.
- Material identification.
- Storage and preservation.
- Handling and rigging.
- Identification, segregation and disposition of non-conforming material.

6.1.2 The Project Manager may establish an inventory control system for equipment and material received at the construction site as required by the project contract documents.

6.2 Construction Process Controls

6.2.1 The inspection program (discussed in Section 6.3) shall include monitoring construction process controls utilized by construction contractors to verify that these controls are implemented in compliance with applicable construction specification requirements. Construction process controls include:

- Clear and complete instructions for performing work functions, appropriate to the complexity and importance of the activities involved.
- Workmanship standards and criteria.
- Control of special processes, such as welding, heat treating, and application of coatings, including qualification of personnel, procedures and equipment.
- Documentation of correct construction sequence and material identification, such as concrete lift release cards, wire pull and termination cards, and mechanical alignment data.

6.3 Inspection Control

6.3.1 The Project Quality Assurance Engineer shall implement construction inspection activities in accordance with project contract and specification requirements. These inspections shall be accomplished using approved inspection and test procedures (as discussed in Section 3.4). Inspection and test results shall be documented, and reviewed by the Project Quality Assurance Engineer.

6.3.2 The Project Quality Assurance Engineer shall document the training, qualification and certification of inspection personnel, in accordance with project contract and specification requirements.

6.3.3 The Project Quality Assurance Engineer shall maintain a system for identifying the inspection status of equipment, systems, and structures subject to construction inspections. Inspection status shall be indicated by stamps, marks, tags or labels attached to the item, or on documents such as drawings, construction travelers, or inspection records traceable to the item.

6.3.4 The inspection procedures shall include monitoring the following construction contractor activities for compliance with project contract and specification requirements:

- Design document control (discussed in Section 4.1).
- Field changes (discussed in Section 4.2).
- Material controls (discussed in Section 6.1).
- Construction process controls (discussed in Section 6.2).

APPENDIX 7B

EXAMPLE: NEW YORK CITY TRANSIT AUTHORITY (NYTA)

Following are some of the forms used as a general conditions appendix to the NYTA quality assurance procedures:

- Concrete Placement Checklist
- Structural Steel Checklist
- Waterproofing Checklist
- Deficient Strength Cylinder Checklist

CONCRETE PLACEMENT CHECKLIST

This checklist is not a substitute for the drawings and specifications. The inspector is responsible for familiarizing himself with the contract documents, ASTM, PCA or ACI Standards or any applicable Transit Authority Procedures.

Placement
Activity No. ______________________ Ambient Temp.________________

Placement Area ________________ Form Temp. __________________

____

Date of Placement ______________

PRE-PLACEMENT	YES	NO	N/A
1) Formwork Preparation			
A) Forms braced properly	( )	( )	( )
B) Forms clean	( )	( )	( )
C) Approved coating material on forms	( )	( )	( )
D) Water stops properly oriented in pour	( )	( )	( )
E) Water stop joints fastened together	( )	( )	( )
F) Construction joints clean	( )	( )	( )
G) Forms sufficiently tight to prevent loss of mortar	( )	( )	( )
2) Reinforcing steel Placement			
A) Bars free of mud, oil, grease or other contaminant	( )	( )	( )
B) Bars fastened together properly	( )	( )	( )
C) Bars placed in accordance with approved rebar drawings	( )	( )	( )
D) Have all embedded items been placed	( )	( )	( )
E) Will all embedded items, including re-bars get complete concrete coverage in accordance with drawings	( )	( )	( )
3) Concrete Batch Ticket received from Truck and properly filled out	( )	( )	( )

4) Received concrete at point of Placement			
A) Proper Mix	( )	( )	( )
B) Contractor ready to place	( )	( )	( )
5) Preplacement Remarks:			

____

SIGNATURE, TITLE

DURING PLACEMENT	YES	NO	N/A
1) Concrete Testing			
A) Slump within tolerance Mass Concrete 1 to 3 inch slump Reinforced Concrete 2 to 4 inch slump Very Constricted Placing Conditions	( )	( )	( )
B) Time Limitation adhered to (1 1/2 hrs from batching to forms)	( )	( )	( )
C) Air Content, within limits (6% ± 2% of the vol. of concrete when tested in accordance with ASTM C173)	( )	( )	( )
D) Concrete temperature within tolerance (50°F to 90°F PCA 12th ED.)	( )	( )	( )
E) Cylinders Taken 9 cylinders for every 150cu. yds. or less for non building code regulated concrete. 9 cylinders for every 60cu. yds. or less for non building code regulated concrete (where no batch plant inspector is available) 9 cylinders for every 50 cu yds. or less for building code regulated concrete.	( )	( )	( )

2) Concrete Placement

A) Means of Placing — Crane & Bucket ( )
— Buggy ( )
*Explain ________ — Chute ( )
________________ — Conveyor Belt ( )
________________ — Pumping ( )
________________ — Other * ( )

	YES	NO	N/A
B) Distance of Concrete Free Fall			
1) Limited to 4 ft maximum without elephant trunk	( )	( )	( )
2) Segregation occurring due to free fall	( )	( )	( )
C) Vibrating Requirements			
1) Backup vibrator on site	( )	( )	( )
2) No horizontal movement of concrete	( )	( )	( )
3) Sufficient vibrating to assure consolidation of concrete	( )	( )	( )
4) Vibrate around reinforcements and embedments	( )	( )	( )
5) Penetrate into previous layer by 6"	( )	( )	( )

	YES	NO	N/A
6) Vibrating in one position for no more than 15 seconds (PSA 12th Edition)	( )	( )	( )

D) Finish
- 1) Brooming ( )
- 2) Rough ( )
- 3) Steel Trowel Finish ( )
- 4) Float Finish ( )

E) During Placement Remarks:

____
SIGNATURE, TITLE

POST PLACEMENT	YES	NO	N/A
1) Method of curing			
A) Date:______________Outside Temp.________			
B) Type of Curing			
1) Ponding or Continuous sprinkling	( )	( )	( )
2) Absorptive material or fabric continuously wet	( )	( )	( )
3) Waterproof sheet	( )	( )	( )
4) Curing Compound	( )	( )	( )
C) Corrective measures required	( )	( )	( )

Temperatures Below 32° or Above 90°

Date________	Date________	Date________	Date________	Date________
Temp. ______	Temp. ______	Temp.________	Date________	Date________

2) Post Placement Remarks:

____
SIGNATURE, TITLE

STRUCTURAL STEEL CHECKLIST

Date __________________
Location of
Steel Erection: _____________

PRE-ORDER & DELIVERY CHECK	YES	NO	N/A
1) Supplier Approved	( )	( )	( )
2) Erection Contractor Approved	( )	( )	( )
3) Has the Material Quality and Reliability Section Approved the Materials	( )	( )	( )
4) Have the specifications for structural steel and the contract drawings been studied	( )	( )	( )

PRE-ERECTION CHECKS

1) Shop Drawings Approved	( )	( )	( )
2) Has the contractor's erection plan and shop drawings been studied	( )	( )	( )
3) Steel stored off ground on timber blocks or skids	( )	( )	( )
4) Has the structural steel brought to the job site been inspected and approved by the Material Quality and Reliability Section	( )	( )	( )
5) Have the proper members been indentified at the job site	( )	( )	( )
6) Has the steel been mishandled incurring bends, twists, kinks or the shop coat been removed. Badly damaged pieces shall be rejected and replaced	( )	( )	( )
7) Has the steel been cleaned of all dirt, loose scale and rust using wire brushes?	( )	( )	( )
8) Has line and grade for pads, grillages base plates been checked	( )	( )	( )

ERECTION CHECKS

1) Is the steel good for line & grade?	( )	( )	( )
2) Is the steel plumb?	( )	( )	( )
3) Correct members for this area?	( )	( )	( )
4) Correct bolts being used?	( )	( )	( )
5) Are washers installed under the turned element (nut or head)?	( )	( )	( )
6) Impact wrench calibrated?	( )	( )	( )
7) Are abutting surfaces clean & parallel?	( )	( )	( )
8) Have all erection bolts been replaced?	( )	( )	( )
9) Torque wrench calibrated?	( )	( )	( )
10) Bolt torque checked? Usually 10% of a pattern is checked,... if any fail, the entire connection is checked.	( )	( )	( )
11) Has steel been braced to prevent movement during erection of concrete forms?	( )	( )	( )

____
SIGNATURE, TITLE

WATERPROOFING INSPECTION CHECKLIST

This checklist is not a substitute for the drawings for the contract drawings and specifications. The inspector is responsible for familiarizing himself with the contract documents or any applicable Transit Authority Procedures.

Date ________________
Location of
Application: ____________

PRE-APPLICATION

Have the following items been checked?	YES	NO	N/A
1) Sub-contractor/contractor Approved	( )	( )	( )
2) Approval of materials by Material Quality and Reliability Section	( )	( )	( )
3) Shop drawings Approved	( )	( )	( )
4) Material brought to site approved (should bear approval tag from MQ&R)	( )	( )	( )
5) Have you checked all details of waterproofing as per contract drawings & Specs.	( )	( )	( )
6) Have manufacturers instructions and application procedure been reviewed.	( )	( )	( )

DURING APPLICATION

Have the following items been checked?			
1) Surfaces to receive Waterproofing smooth, dry and clean	( )	( )	( )
2) Proper Temperature	( )	( )	( )
3) Proper amount of relative humidity	( )	( )	( )
4) Thickness of material or number of layers to be applied	( )	( )	( )
5) Proper time for curing between applications	( )	( )	( )
6) Surface dry and clean between applications	( )	( )	( )
7) Treated surface protected	( )	( )	( )

PRECAUTIONS DURING APPLICATION

a) Material is flammable. Smoking and open flames of any type should not be permitted in the vicinity of applications	( )	( )	( )
b) Prolonged inhalation of vapors should be avoided. In closed areas, provide proper exhaust system for fumes.	( )	( )	( )
c) Protective clothing should be worn during application	( )	( )	( )
d) No loads to be placed on exposed waterproofing	( )	( )	( )
e) Exposed waterproofing must be protected, i.e., layer of mortar, plywood, etc.	( )	( )	( )

____

SIGNATURE, TITLE

DEFICIENT STRENGTH CYLINDER CHECKLIST

Concrete Placement Location______________________________

Steps to take to resolve problem:	YES	NO	N/A
1) Identify area of placement	( )	( )	( )
2) Project Manager to make request for In-Place test	( )	( )	( )

3) In-place tests performed
   a) Concrete Passes ( ) Prepare report
      No further action required
   b) Concrete fails ( ) Request designer input to evaluate design strength.
      Project Manager's office provides input to Deisgner i.e. extra thickness of concrete in area, embedded temporary steel that wasn't removed etc. Project Manager directs Contractor to make cores of placement area for testing by Materials Quality and Reliability section.
4) Materials Quality and Reliability Department test cores.
   a) Concrete Passes ( ) Prepare report
      No further action required.
   b) Concrete fails ( ) Project Manager reviews with Designer the results of their analysis.
      1) Analysis of structural integrity indicates that reduced strength concrete is acceptable ( ) Project Manager prepares report. No further action required.
      2) Analysis of structural integrity indicates unacceptability of reduced strength concrete ( ) Contractor is directed to remove deficient area of concrete at his own expense. ( ) Project Manager prepares report. No further action required.
      3) Contractor refuses to remove deficient area ( ) Project Manager applies Procedure for Deviations and non-conformance.

CHAPTER 8

Documentation of Field Administration

MOBILIZATION

§ 8.1 At the start of a project, contractors selected by competitive bid must mobilize their forces. This initial period of organization is also one of organization for the inspection staff. Usually, it is good practice for the construction management resident team, including the field engineers, to conduct a job conference. This initial conference reviews the steps the contractors will take at the beginning of the project to get organized and to respond to contractual commitments.

These contractual requirements include, among other things, the performance bonds, liability insurance by the contractors, equal opportunity compliance documentation, building permits, and other special requirements of the project. In addition, the contractor at this stage submits initial material on the job schedule, payment schedule, shop drawing submission plan, and materials submission schedule. Special information such as the trial concrete mix should also be submitted during this mobilization phase.

The supplemental conditions of the contract may require the contractor to furnish facilities for the inspection team if the site is in a remote location. The initial meeting will review such factors as well as needs such as temporary heat for the building and other special requirements of the job.

BONDING

§ 8.2 It is usual practice for the owner to require the contractor to submit a bid bond with the bid. The purpose of this bond, issued by the contractor's bonding company, is to ensure that if the contractor is awarded the job he or she will perform the work. Further, once a contractor is selected, it is common practice for an owner to require the contractor to post a completion bond which permits the owner to complete the work. Should the contractor default, the bonding company will arrange to complete the work by providing the funding for this contractor or another one to carry out the work to conclusion.

Just as the borrower who has sufficient collateral—and therefore less real need to borrow—is especially well received at the bank, so too, is the contractor favored who causes the least concern regarding his or her ability to complete the work. The successful contractor, stable in the financial sense and sound of business reputation, has less difficulty in receiving a bond than the less well-known or more financially unstable contractor.

The bonding company reviews the contractor's financial statements, work experience, backlog of current work, credit references, and current status with material suppliers and subcontractors before agreeing to provide either bid or completion bonds. The fee for bonds is a function of the risk involved, and ranges from less than one percent to several percent, depending upon the situation. The bonding company is interested in the cumulative amount of work in progress, and the contractor's ability to be bonded in cumulative terms is a direct reflection of the reputation, financial standing, and general well-being of the contractor.

The contractor may also be required to supply a special bond to suppliers, subcontractors, and others. Generally, the greater the bond required, the less stable the contractor.

In difficult relationships with contractors, it is not unusual for the owner to threaten to turn the job over to the bonding company. This is more easily said than done, however, as the bonding company specializes in not taking over jobs. In any confrontation, the inspection team can anticipate that the bonding company will be very much on the side of the contractor, as it is the bonding company's preference that the contractor finish the job on the contractor's funds and terms. While the bonding company is committed to ensure that the contractor in default will be replaced or assisted in the completion of the building, it is not unusual, even in cases of bankruptcy, for a bonding company to keep the same contractor on the job. Bonds provide some protection, but they do not generally ensure that a building will be completed in a timely fashion. Completion also becomes a negotiable area in cases of default. Further, even in cases in which the bonding company does agree to a hard and fast completion date, that date can be shifted or negated by changes in scope or changes in conditions beyond the control of the owner or the contractor.

INSURANCE

§ 8.3 The general conditions of the contract establish the level of insurance that the contractor must carry. The owner's principal interest is to avoid any liability that may pass through an underinsured contractor to the owner. Today, it is quite usual for owners and other parties to the contract, as well as the contractor, to be sued for injuries, accidents, damages, and other mishaps on a project even when they occur within the physical and contractual limits of the contractor's field of responsibility. The inspection team or the owner's office staff should periodically check on the validity of the insurance—specifically that it has been neither cancelled nor allowed to lapse.

Generally, as a matter of state law, the contractor must agree to standard worker's compensation and employer's liability policies. There are variations in different states, and the owner's reviewing personnel must be aware of the requirements of the state in which the project is being built.

Beyond the minimal statutory worker's compensation and employer's liability, most contracts require the contractor to carry a comprehensive general liability policy. Typically, the general liability policies are broken down into five divisions:

- Operations on premises: covers bodily injury or property damage caused by occurrences on the premises owned or leased by the insured anywhere in the country, or as limited by the policy.

- Elevator liability: covers legal liability for injury or property damage by occurrences involved in the ownership, maintenance, and use of elevators that are controlled and operated by the insured, usually not including material hoists for construction operations.
- Protective liability: covers the contractor's legal liability for bodily injury or property damage arising out of operations performed for another. Policies usually include provisions to cover secondary liabilities resulting from sublet operations or subcontracts.
- Completed operations: covers the protection of the completed operation, and to some degree repeats operations-on-premises coverage. Usually, it is the contractor's responsibility to protect the contractor's completed work, but often the experienced owner will require this additional protection to preclude the necessity for litigation where serious vandalism or damage arises on a project, and the contractor is unable financially to repay the difference.
- Contractual liability: covers contractor's liability for bodily injury or property damage that is assumed under written contract with other parties. This would cover incidental subcontracts, or services retained by the contractor and not coming under operations-on-premises coverage.

The owner's interest in specifying insurance coverage on the part of the contractor lies solely in the protection of the owner from liability or litigation. Accordingly, specification for coverage is a consideration of the owner's legal counsel, and subsequent coverage provided by the contractor should be reviewed by the legal counsel for the owner. The work of the inspection team in this area is limited, but the owner's inspection team should be aware of the coverage enforced. Further, in the event of any accidents involving damage or bodily harm, the inspection team becomes the owner's on-site representatives and should carefully investigate situations, functioning to mitigate difficult circumstances for all parties. However, the inspection team's first responsibility is to the owner, and care must be taken not to involve the owner in problems which properly should be limited to the contractor.

EQUAL OPPORTUNITIES

§ 8.4 Most requirements for equal opportunity employment use as a minimum base the Civil Rights Act of 1964, which provides that it is a violation of federal law for an employer employing more than 24 people to discriminate in hiring, business operations, or layoffs with respect to race, color, religion, sex, or national origin. On federal work, the contractor as part of the bid must assure that all equal opportunity regulations covering construction work as described by the executive order of the president, and enforced by the Office of Federal Contract Compliance of the Department of Labor, will be met.

Although it is the contractor's responsibility to meet the requirements of the Civil Rights Act, the inspection team is responsible for the collection of information on the actual job work force, including compliance. The inspection team should be familiar with the current requirements of the executive order and local requirements of the Office of Federal Compliance, often described in terms of the locale, such as the New York Plan, the Philadelphia Plan, and so on. The attachment of labels such as these is sometimes misleading, since plans and practices do shift.

In addition to federal law, individual states often have specific civil rights laws and programs. Because these are statutory obligations, failure to comply with them can result in delays and even in enforcement passing through the contractor to the owner.

LABOR RELATIONS

§ 8.5 Relations with the trades are the responsibility of the contractors, but the inspection force can have an influence. Many projects involving inspection are either wholly or partially unionized, and a knowledge of the local working conditions and agreements is important to the inspection function. The field engineer is, after all, the construction manager's eyes and ears and has the responsibility to forecast problems in all areas, particularly those involving labor. The building trades comprise more than 20 crafts, which in turn are broken down into many specialties. Often, problems develop over jurisdiction between trades, in some cases triggering work delays or stoppages in which neither the contractor nor the owner are directly involved, but from which both directly suffer.

Usually, the working agreements specify that the inspection team must communicate with the trades via their supervisory structure. This is good management; informal liaison often leads to problems. In the working agreements, there are even differences for a given craft between local union agreements and international agreements.

The inspection team must assume a laissez-faire posture in the relationships between the contractor and the labor force, carefully maintaining the role of passive observers. It is not the prerogative of inspection team members to be either for the contractor or for labor; they are only for their client, the owner.

On-the-job disagreements that escalate may have to be settled by the national joint board for the settlement of jurisdictional disputes for the National Appeal Board. Prior to this, the National Labor Relations Board has jurisdiction in attempting to settle any jurisdictional disputes between unions. However, by the time these rather ponderous procedures are put into action, harmful delays will have taken place. The best place to settle jurisdictional disputes is at the local level, before they become acute enough to cause walk-offs.

BUILDING PERMITS

§ 8.6 The contract often calls upon the contractor to obtain the building permit; in the case of multiple prime contracts, this is usually the responsibility of the general contractor. Naturally, where there are defects or omissions in the plans and specifications, it is the owner's responsibility to take remedial action.

Building codes may be local, city, county, or state. There is a recommended national building code, but it usually has no effect, except on federal projects. In turn, federal projects built on state, city, or county lands, either public or private, are considered to be essentially a federal reservation, and local laws do not apply.

The regulations and subsequent review of building codes involve categories within the building such as exits, egress and access, fire protection, setbacks or building lines, building lines, building classifications, allowable stresses and loads, foundations, structure, heating, ventilating and air conditioning, lighting requirements, noise control, vents and piping, fire alarms, elevators, and any safety factors, particularly if the building will be used for public assembly. Often, even though a pre-building permit or preliminary review has been conducted by the design team, or a permit to allow construction has been awarded, the builder must get separate permits allowing work in certain categories such as foundations, demolition, plumbing, equipment use, and signs.

Perhaps even more important in many areas is the requirement for an occupancy certificate before the building may be utilized. Often, the owner uses this as a condition of completion and final payment. In many cases, a beneficial occupancy certificate may be issued as a matter of normal practice; and in some cities where this is the case, final building occupancy certificates are rarely issued.

The building department also often handles the licensing in the locality. Local regulation often calls for electricians and plumbers in particular to be licensed.

JOB MOBILIZATION CHECKLIST

§ 8.7 Material submitted should include, but not be limited to:

- appropriate insurance certificates;
- performance bond;
- equal opportunity documentation;
- building permits;

- special permits;
- progress schedule;
- payment schedule;
- shop drawing submission plan;
- materials submission plan;
- concrete mix design;
- subcontractor list; and
- emergency telephone number list.

PROGRESS PAYMENTS

§ 8.8 An important function of the inspection team is the approval of progress payments. For the contract that is progressing at an appropriate pace, with a good attitude on the part of the contractors, the expeditious handling of progress payment requisitions results in a fast return of invested money to the contractor, and thus lower interest costs. On a project that is not going well, the approval of progress payments is one of the best ways for the inspection team to maintain control and keep the contractor's interest.

There can be legal questions about overpayment. A contractor may receive progress payments too far beyond the actual work. Should the contractor choose to accomplish or be forced to default, the contractor would leave the owner in jeopardy of losing the overinvestment or overpayment.

Figure 8-1, "Schedule of Values," allows for allocation of the total contract price. Figure 8-2, "Detailed Breakdown," itemizes the costs for a work activity and lists previous and current payments. Figures 8-3 and 8-4 are forms submitted by the contractors when requesting monthly progress payments. Figure 8-5 assures that the pay requests are forwarded to all involved parties. Figure 8-6 is used to itemize the cost of extra work authorized by the owner.

SCHEDULE OF VALUES
(Allocation of Total Contract Price)

PROJECT TITLE		PROJECT NO.		SPEC. OR CONTRACT NO.	PAGE OF
NAME OF CONTRACTOR		CONTRACT TYPE		CONTRACT DATE	
SCHEDULE & ITEM	DESCRIPTION OF COMPONENT PARTS	ALLOCATION*		TOTAL	
		MATERIAL	LABOR		
	TOTAL			**	

* Allocation shall be in percents of total contract price of the items allocated or in dollars and cent

** This total shall be equal to the total contract price of all items allocated.

Figure 8-1

DETAILED BREAKDOWN

PAGE ______ OF ______

CONTRACT NO.	PROJECT	REPORT NO.	DATE

ITEM OF WORK (1)	TOTAL VALUE OF WORK (DOLLARS ONLY) (2)	VALUE OF WORK COMPLETED (3)		TOTAL VALUE OF COMPLETED WORK (4)
		TO LAST REPORT (DOLLARS ONLY) (A)	SINCE LAST REPORT (DOLLARS ONLY) (B)	

Figure 8-2

REQUEST FOR PAYMENT

CONTRACTOR		PAGE ______ OF ______	
LOCATION			
PROJECT TITLE		REPORT NO.	DATE
CONTRACT NO.	PROJECT NO.		

WORK STATUS:

COMPLETION DATES				PERCENT COMPLETED	
INITIAL CONTRACT	REVISED CONTRACT	ESTIMATED SUBSTANTIAL	ACTUAL SUBSTANTIAL	THROUGH THIS MONTH (SUM OF LINE 4 AND LINE 5 + LINE 3)	NORMAL TO DATE

AVERAGE WORK FORCE	PROGRESS	MATERIAL DELIVERIES
NUMBER EMPLOYED	SATISFACTORY ☐ YES ☐ NO	SATISFACTORY ☐ YES ☐ NO
CONSTRUCTION EQUIPMENT	**SHOP DRAWING SUBMISSION**	**SAMPLE SUBMISSION**
SATISFACTORY ☐ YES ☐ NO	SATISFACTORY ☐ YES ☐ NO	SATISFACTORY ☐ YES ☐ NO

REPORT BELOW ANY CIRCUMSTANCES WHICH MAY HAVE ADVERSELY AFFECTED THE PROGRESS SUCH AS STRIKES, WEATHER, DELAYS BY THE OWNER, ETC. INCLUDING EXPLANATION OF ANY "NO" ANSWERS GIVEN IN THE BLOCKS ABOVE.

PROGRESS PAYMENT SUMMARY		
1. INITIAL CONTRACT AMOUNT		
2. CHANGE ORDERS (Total of Column 2, Form 8-D)		
3. TOTAL CONTRACT AMOUNT TO DATE (Line 1 plus Line 2)		
4. VALUE OF WORK COMPLETED TO DATE (Total of Columns 3A and 3B of Form 8-C)		
5. VALUE OF WORK COMPLETED UNDER CHANGE ORDERS (Total of Column 4, Form 8-D)		
6. VALUE OF MATERIAL		
A. MATERIAL ON SITE		
B. MATERIAL IN STORAGE		
7. TOTAL VALUE OF MATERIALS (Line 6A plus Line 6B)		
8. TOTAL VALUE OF COMPLETED WORK AND MATERIALS (Sum of Lines 4,5 & 7)		
9. LESS RETAINAGE		
10. SUB-TOTAL (Line 8 minus Line 9)		
11. LESS PREVIOUS PAYMENTS		
12. AMOUNT OF PAYMENT THIS REPORT (Line 10 minus Line 11)		

SIGNATURE	DATE	SIGNATURE	DATE
SIGNATURE	DATE	SIGNATURE	DATE

Figure 8-3

REQUEST FOR PAYMENT FOR MATERIALS ON HAND

INSTRUCTIONS:

TO CONTRACTORS:

Forward original and one copy to Resident Project Representative. Attach evidence of purchase (and warehouse receipt when required) to the original.

TO RESIDENT PROJECT REPRESENTATIVE:

Retain original in your files with supporting documents for progress payments.

TO (RESIDENT PROJECT REPRESENTATIVE)	DATE
	PROJECT NO.
FROM (CONTRACTOR)	CONTRACT NO.

In accordance with the provisions of the General Conditions of the Contract, request is made for payment of materials on hand for the following materials:

ITEM NUMBER	QUANTITY	UNIT	MATERIAL DESCRIPTION	VALUE	WHERE STORED

AFFIDAVIT

The materials listed above have been purchased exclusively for use on the above referenced project. The material is separate from the other like materials and is physically identified as our property for use only on Contract No.__________. The Owner may enter upon the premises for inspection, checking or auditing, or for any other purpose as you consider necessary. It is expressly understood and agreed that this information and affidavit is furnished to the Owner for the purpose of obtaining payment for the above materials before they are delivered to, or incorporated into, the project described above, and that the storage thereof at the location shown shall not relieve the Contractor of full responsibility for the security and protection of all such materials until acceptance by the Owner of the completed project.

CONTRACTOR: BY______________________ TITLE______________________ DATE______________

Figure 8-4

CONTRACT PAYMENT ROUTING SHEET

CONTRACTOR		
CONTRACT NUMBER	INVOICE NUMBER / DATE	

ACTION	INITIAL	DATE
1. PAYMENT REQUEST RECEIVED BY CONSTRUCTION MANAGER		
2. REVIEW BY ASSISTANT CONSTRUCTION MANAGER COMPLETED		
3. SIGNED BY ASSISTANT CONSTRUCTION MANAGER		
4. SIGNATURE BY CONSTRUCTION MANAGER		
5. RECEIVED BY OWNER		
6. SIGNATURE BY OWNER		
7. RECEIVED BY OWNER'S ACCOUNTS PAYABLE		

COMMENTS

Figure 8-5

CHANGE ORDER BREAKDOWN

PAGE ______ OF______

		REPORT NO.	DATE
CONTRACT NO.		PROJECT	

		VALUE OF WORK COMPLETED (3)		
CHANGE ORDER NO. (1)	TOTAL VALUE OF WORK (DOLLARS ONLY) (2)	TO LAST REPORT (DOLLARS ONLY) (A)	SINCE LAST REPORT (DOLLARS ONLY) (B)	TOTAL VALUE OF COMPLETED WORK (4)

Figure 8-6

Progress Payment Basis

§ 8.9 The basis of progress payments is, in all cases, an estimate of the work actually accomplished. This assessment may be determined by actual field observation of weights and quantities, by acceptance of contractor statements, or by a combination of the two. The method and accuracy of work measurement varies with the type of contract.

There are many variations of construction contracts, but most break down into one of the following: (1) fixed price, lump sum; (2) fixed price, per unit quantity; and (3) cost plus fixed fee.

In the lump-sum contract, which is usually competitively bid with sealed confidential bids, the contractor selected is usually the one with the lowest responsive bid price. The contract is qualified only by the fact that the contractor must be shown to be a responsible individual, as proven by both the contractor's record and ability to be bonded for the amount of the bid. Shortly after the acceptance of the successful bid, the contractor is required to submit a cost breakdown. This is a detailed breakdown of the project costs by specified categories. Usually, the owner details the major categories, and the contractor has the flexibility to select subcategories. The purpose of the project breakdown is to facilitate approval of progress payments.

Usually, the designer and/or construction manager assists the owner in the approval of the job breakdown. In some cases, the inspection team may also have the opportunity to participate. The owner and the owner's representatives are trying to avoid front-end loading of the payment schedule, which is a fairly general practice whereby the contractor weighs the early activities more heavily in order to pull the investment out of the job as it progresses. The owner counters this practice by careful review of the job cost breakdown. Another counter is the fact that traditionally payments are on a monthly basis, and, therefore, the contractor has invested that month's work plus whatever approval on payment cycle time lag the owner imposes, which is usually a month or more.

The more detailed the breakdown, the more readily it can be reviewed. For cost-loaded schedules, the cost breakdown is keyed to the schedule. Where the critical path method is used, the breakdown should represent the cost by activity. This not only avoids keeping two sets of books, one for time and the other for cost, but also avoids out-of-sequence claims for progress payments. (Without the critical path method sequence information, such claims may be made inadvertently, or in a few cases, deliberately.)

In many heavy construction jobs, such as major highways, lump-sum unit price bids are often taken. A typical bid of this type presents a categorical breakdown of units, such as excavation—earth; excavation—rock; backfill—gravel; and paving per cubic yard. Perhaps 25 to 50 categories are specified, and estimates are given of the amounts for each category. The contractor, then, presents a lump-sum estimate, but specifies a set fee per unit. The actual work is measured, and payment is made upon the quantities of work accomplished, regardless of the quantities listed by the designer. If a careful quantity survey has been done by the designer, and too many unforeseen conditions do not occur, the final price should be close to the quantities upon which the bid was based.

In responding to a request for this type of bid, the contractor may apply strategy in the cost breakdown. Naturally, the unit costs bid, when extended by the quantities listed by the designer, result in a fixed price for terms of comparison only. However, the contractor may utilize his or her own estimation and cost information to arrive at a total proposal price, and then may revise the units in anticipation of actual job conditions. Knowing that the bid evaluators will review the unit prices (which become the basis for adding or deducting from the base fee in direct proportion to the actual quantities of work), the contractor may attempt to determine where the designer may be over- or underestimating quantities. One tactic is to present a very favorable unit price where the contractor thinks there really will be less cost in a category than is listed. For those categories where the contractor expects substantially greater amounts of work, he or she may quote a unit price favorable to the contractor. This will result in a higher profit margin should the quantities run over the base estimate of the owner.

The inspection team has little to do with the establishment of prices in this type of contract. However, it plays a vital role in the determination of the actual quantities of work accomplished. Payment is on the basis of units of count, length, area, volume, weight, or, for some items, lump sum. The unit of measurement for any particular category is given in the specifications and the unit breakdown list. In the field, quantities are measured and calculated to a degree of accuracy consistent with the value of the item.

Quantities of some items, such as paving, fencing, and sewer lines, are measured in terms of their size and length or area. Measurement of volume by vehicle loads is not a recommended method of determining quantity because loading factors vary with the equipment, weather, and operators. Usually, excavation or fill is measured by a field survey or by weight on a truck scale. If it is done by load, a bulking factor of 15 to 20 percent should be deducted.

When work has been paid for in a previous estimate and loses value through damage, loss, theft, or failure to function, the value lost should be deducted from the following monthly requisition.

In certain cases, the owner's regulations permit payment for materials delivered to the job and stored for future installation. Generally, only a set percentage of the value of the material is paid as a progress payment.

In some instances, the special provisions of the specifications list specific materials eligible for inclusion in progress payments when delivered to the job site. The inspection team should carefully check over any materials claimed for payment when they are allowed under the special provisions of the specifications. If this is not done, it is quite possible that nonspecified material, or even incorrect material, may be delivered to the job site and paid for. The inspection team also should be concerned with the methods of unloading, as well as storage facilities, when a material is sensitive to handling or the elements. Security is also a consideration; the contractor is usually required to provide fencing, guards, and other security measures.

In some cases, credit may be given when the material is shipped to a specified warehouse or holding area. The inspection team should make a point of verifying the delivery of major items.

INDEX CATEGORIES

§ 8.10 The success of the documentation control system rests on the filing system, correspondence control, and technical control logs. Each project should have three basic types of files: program files, contract files (design and construction), and administrative files.

Program Files

§ 8.11 The program or project files will contain information relating to the program as a whole but will exclude information relating to specific contracts (both design and construction) and the on-site office administration files. The first category within the program administration file should be the chronological file, which will contain original copies of all incoming correspondence and copies of all outgoing correspondence by date. In addition, there should be categories relating to financial information, including budgets, forecasts, and estimates; general schedule data, including the master schedule and occupancy schedules; procurement information for both design and construction contracts; and the purchase order file relating to the project as a whole. There should also be areas for design, bidding, construction, meetings, reports, and issues.

Administrative Files

§ 8.12 Administrative files are those required to operate the job site office and will contain information on such items as time and attendance, the contract, purchasing, office equipment, staff planning, and the home office.

Contract Files

§ 8.13 Contract files can be broken down into two basic types: those for design and those for construction. The design file should include areas for the architect/engineer contract and major portions of the design process, such as feasibility, schematics, design development, working drawings, reports, meetings, schedules, and estimates. All these issues should relate directly to the specific architect/engineer contract.

Files maintained for the construction contract should include areas for the contract, specifications, drawings, labor, safety, meetings, reports, schedules, payments, submittals, requests for information, field instructions, changes, claims, inspection and testing, photographs, owner-furnished materials, coordination, closeout and acceptance, occupancy and transition, and issues.

File Codes

§ 8.14 The three major files should be set up with a file index code. A sample system follows:

1. The program or project file will be given a letter designation "A" followed by the numerical subcategory designations.
2. The administrative files will be given a letter designation "B" with the related numerical subcategories.
3. The contract file codes will be set up with the remaining letter designations "C" through "Z" as required with the numerical subcategory designation.

Filing Process

§ 8.15 Basically, two types of files must be maintained for all documents: chronological and functional. All incoming correspondence should be date stamped and receive the appropriate chronological file code, plus the appropriate functional file code for the program, administration, or contract file. There will be at least one original and one copy of all incoming documents filed in the chronological and functional files. Incoming correspondence should be given a chronological file code immediately upon receipt and the project manager should assign the additional functional codes.

Logs

§ 8.16 In addition to being coded for the appropriate file, all incoming and outgoing correspondence will be controlled by correspondence control logs. Logs may be kept manually or computerized. Figures 8-7 and 8-8 show examples of manual incoming and outgoing correspondence logs. Each incoming or outgoing document will be provided with a sequential incoming or outgoing number. The log will contain the code number, the date received, and the subject. A more sophisticated log will reference the specification, the response due date, the actual response date, the response letter reference number, and the file number.

In addition to the file system and correspondence control system, there should be a number of technical control logs. These logs can be either manual or computer generated, depending on the size of the project. The following is a partial listing of technical logs which should be used on a project.

Request for Information (RFI) Control Log

§ 8.17 This log is the heart of the clarification, change order, and dispute process. The log contains all questions from the contractor and clarifications and bulletins generated by the construction manager or architect/engineer. Many of the answers will be considered changes or disputes which should be cross-referenced.

Proposed Change Order (PCO) Log

§ 8.18 This log follows the change order process from the request for a price quote from the contractor through negotiations and finally to the approved change order. This is cross-referenced with the request for change, field instruction, and request for information logs.

Request for Change (RFC) Log

§ 8.19 This log may be based on the proposed change log. It tracks requests for changes initiated by the contract or

INCOMING CORRESPONDENCE LOG

CORRESPONDENCE NUMBER	LETTER DATE	COMPANY NAME/ SYMBOL	SUBJECT OF CORRESPONDENCE	SUBJECT CODE
1/				
1/				
1/				
1/				
1/				
1/				
1/				
1/				
1/				
1/				
1/				
1/				
1/				
1/				
1/				

Figure 8-7

OUTGOING CORRESPONDENCE LOG

CORRESPONDENCE NUMBER	LETTER DATE	COMPANY NAME/ SYMBOL	SUBJECT OF CORRESPONDENCE	SUBJECT CODE
O/ .LTR				
O/ .LTR				
O/ .LTR				
O/ .LTR				
O/ .LTR				
O/ .LTR				
O/ .LTR				
O/ .LTR				
O/ .LTR				
O/ .LTR				
O/ .LTR				
O/ .LTR				
O/ .LTR				
O/ .LTR				
O/ .LTR				

Figure 8-8

whether they have been answered, and whether a change order was issued or the request was rejected. The RFC log should be cross-referenced to the proposed change, request for information and dispute logs.

Submittal Log

§ 8.20 The submittal log tracks the approval process of submittals to and from the contractor.

Claims Log

§ 8.21 The claims log will list all potential and actual disputes concerning requests for time or money. It will include requests for change which have been rejected, construction manager final decisions, and formal disputes.

Field Instruction (FI) Log

§ 8.22 This log includes a listing of all directives given on a field instruction form.

Notice of Noncompliance Log

§ 8.23 This log is a listing of all deficiency notices generated during the project until they have been completed by the contractor.

Hold Order Log

§ 8.24 To the extent that access to any portion of the project is denied to the contractor, a log describing the hold, the date issued, who issued the hold, and the release date should be maintained.

PROGRESS MEETINGS

§ 8.25 A regularly scheduled weekly progress meeting should be established at the preconstruction conference. Attendees should include the construction management staff, the contractor, and representatives of each major subcontractor, as appropriate to the project's progress. Figure 8-9 is an example of a meeting attendance sign-up sheet for accurate tracking of attendees.

The project manager or field engineer should prepare an agenda (which usually is derived from the previous meeting's

MEETING ATTENDANCE

MEETING TYPE:			NO.
DATE	TIME	LOCATION	

ATTENDEES

NAME	AFFILIATION	PHONE

Figure 8-9

minutes) for discussion at these meetings. The agenda should include a list of outstanding items which will be reviewed as appropriate. A suggested list of ongoing categories for discussion would include:

- submittals/shop drawings;
- requests for information;
- field instructions;
- payments;
- schedule/progress;
- force account work;
- progress payments;
- safety and security;
- change orders;
- claims; and
- quality control.

Problems on the job are then broken down by category and assigned corresponding log numbers. In addition to the ongoing categories of discussion listed above, time should be reserved to review any unresolved issues. These may be brought up by the representatives attending the meeting. When an issue cannot be resolved at the meeting in which it was first discussed, the project manager will assign it a two-part consecutive log number for future tracking of the problem and its resolution. The first number should identify the meeting at which the problem was discussed; the second should be a consecutive issue number.

Minutes should be taken by the project manager at all weekly progress meetings and special meetings. The minutes should clearly identify all unresolved issues discussed, actions agreed to, and who is responsible for taking those actions. Minutes should be distributed to all attendees in a timely manner and reviewed prior to the next regularly scheduled meeting. Appendix 8A shows an example of progress meeting minutes.

OTHER MEETINGS

§ 8.26 During the course of the project it will be necessary to schedule additional meetings to review specific issues on topics which are too involved for inclusion in the weekly progress meeting. When a special meeting is required, the project manager will coordinate the time and place for all required attendees.

SHOP DRAWINGS

§ 8.27 The monitoring of shop drawings is a field administrative duty of vital interest to quality control. It is essential that the quality control inspection team have a current list of approved shop drawings available to ensure that the work being installed, including any changed work, is based upon the latest approved drawings. On major projects, the shop drawing list should be computerized.

Who Should Keep the Log

§ 8.28 There are several options regarding who will set up and maintain the shop drawings log:

- Construction Manager. If the project has a construction manager with a support staff, the log can be set up and maintained by the support staff. This is the optimum method.
- Architect/Engineer. Typically, the architect/engineer maintains a shop drawing log. The field engineer can use the architect/engineer file, which should be kept current to fulfill the primary purpose of identifying the latest approved drawings for the work and/or indicating that a change is in progress, and the previously approved version should not be implemented.

  While this method is convenient for the field engineer, reliance on the architect/engineer file does not fulfill a secondary use of the shop drawing file, which is monitoring the timeliness of the shop drawing processing by the architect/engineer. This is an area which can be important to claims avoidance. If the architect/engineer maintains his or her own file, the architect/engineer can reduce the length of time required for review.

- Contractor. The contractor often maintains a shop drawing file. It usually is not available to the field engineer. It is typically based upon date of submittal and receipt of approval.
- Field Engineer. If a log is not otherwise available, the field engineer can maintain the log manually or on computer.

Computerized Shop Drawing Log

§ 8.29 A shop drawing log is, necessarily, a "living document." Maintenance on either a word processor or computer is an obvious advantage for the following reasons:

- Reports can be generated in hard copy periodically or as needed.
- The file can be sorted by specification section, date of receipt, date of approval, or other key.
- Additional information can be added as received without having to recreate the entire file.
- The file can be used to monitor architect/engineer review time.

There are software packages available for shop drawing files. For smaller projects, the file can be set up on a personal computer with a spreadsheet program such as Lotus 1-2-3 or dBase.

When shop drawing control is based on a computerized log, (see Figure 8-10), each shop drawing record should contain nine pieces of information: the specification section, an assigned computer identification number, the shop drawing, the contractor's transmittal number, a submission number, a description, a transmit-to-engineer date, a transmit-from-engineer date, and an action. Most of this information is taken directly from the contractor's submittal. Figure 8-11 shows that the information may be found on the transmittal form,

REPORT NO 1
LIST BY SHOP DWG —WITHOUT ERRORS

CAMDEN COUNTY MUNICIPAL UTILITIES AUTHORITY
PREPARED BY: O'BRIEN–KREITZBERG & ASSOC.,INC.

PAGE 145
DATA DATE

PROJECT: DELAWARE NO. 1 WPCF CONTRACT: 160 CONTRACTOR: PAUL A. LAURENCE ENGINEER GREELEY & HANSEN

SPEC SECT	ID NO	SHOP DRAWING	TRANS NO.	SUB NO.	DESCRIPTION	TRANSMIT TO ENG	TRANSMIT FROM ENG	ACTION		OUTSTANDING CNTR	SHOPDRAWING ENG
0043	0001	SCHEDULE	0098	01	TOILET ROOM ACCESSORIES	12 18 84	1 10 85	A FYI	(5,7)		
0043	0002	SPECS	0098	01	PAPER TOWEL DISPENSER	12 18 84	1 10 85	A FYI	(5,7)		
0043	0003	SPECS	0098	01	RECESSED ALL-PURPOSE CAB	12 18 84	1 10 85	A	(4,7,8)		
0043	0004	SPECS	0098	01	NAPKIN DISPENSER	12 18 84	1 10 85	AAN-R	(3,7,9)		
0043	0004	SPECS	0098C	02	DUAL NAPKIN/TAMPON DISPNS	3 13 85	3 25 85	A	(4,7)		
0043	0005	SPECS	0098	01	NAPKIN DISPOSAL	12 18 84	1 10 85	AAN-R	(3,7,9)		
0043	0005	SPECS	0098C	02	SANITARY NAPKIN DISPOSAL	3 13 85	3 25 85	A	(4,7)		
0043	0006	SPECS	0098	01	INTER-LOK MIRRORS	12 18 84	1 10 85	AAN-R	(3,7,9)		
0043	0006	SPECS	0098C	02	INTER-LOK SS FRMED MIRROR	3 13 85	3 25 85	A	(4,7)		
0043	0007	SPECS	0098	01	SOAP DISPENSER LAY-BASIN	12 18 84	1 10 85	A	(3,7,8)		
0043	0008	SPECS	0098	01	STAINLESS STEEL SOAP DISH	12 18 84	1 10 85	A	(3,7,8)		
0043	0009	SPECS	0098	01	SHOWER CURTAIN ROD	12 18 84	1 10 85	AAN-R	(3,7,9)		
0043	0009	SPECS	0098C	02	SHOWER CURTAIN ROD	3 13 85	3 25 85	A	(4,7)		
0043	0010	SPECS	0098	01	SHOWER CURTAIN ROD FLANGE	12 18 84	1 10 85	R & R	(2,7,10)		
0043	0010	SPECS	0098C	02	SHOWER CURTAIN ROD FLNGES	3 13 85	3 25 85	A	(4,7)		
0043	0011	SPECS	0098	01	GRAB BAR	12 18 84	1 10 85	A	(4,7)		
0043	0011	1	0316	02	GRAB BARS	3 13 85	3 22 85	A	(4,7,8)		
0043	0012	SPECS	0098	01	SINGLE ROBE HOOK	12 18 84	1 10 85	R & R	(2,7,10)		
0043	0012	SPECS	0098C	02	SS SINGLE ROBE HOOK	3 13 85	3 25 85	A	(4,7)		
0043	0013	SPECS	0098	01	TOILET PAPER HOLDER	12 18 84	1 10 85	A	(4,7,8,11)		
0043	0014	SPECS	0098	01	DUAL ROLL TOILET PAPER	12 18 84	1 10 85	R & R	(2,7,10)		
0043	0014	SPECS	0098C	02	DUAL TOILET TISSUE HOLDER	3 13 85	3 25 85	AAN-R	(3,8)		
0043	0014	SPECS	0098E	03	DUAL TOILET TISSUE HOLDER	2 20 87	3 10 87	A	(4,7)		
0043	0015	SPECS	0098	01	TWIN HIDE-A-ROLL	12 18 84	1 10 85	A	(3,7,8)		
0043	0016	SPECS	0098	01	LIQUID SOAP DISPENSER	12 18 84	1 10 85	A	(3,8)		
0043	0017	SPECS	0098	01	SHOWER CURTAIN	12 18 84	1 10 85	AAN-R	(3,7,9)		
0043	0017	SPECS	0098C	02	SHOWER CURTAIN	3 13 85	3 25 85	A	(4,7)		
0043	0018	C9513	0147	01	TOILET PARTITION DRAWING	1 10 85	2 13 85	R & R	(3,7)		
0043	0018	C-9513	0147A	02	TOILET PARTITIONS	4 29 85	5 10 85	AAN-R	(3,7)		
0043	0018	9513	0147B	03	TOILET PARTITIONS	8 22 85	9 10 85	A	(4,7)		
0043	0019	SPEC	0147	01	TOILET PARTITION SPEC	1 10 85	1 10 85	R & R	(3,7)		
0043	0019	SPECS	0147B	02	TOILET PARTITIONS	8 22 85	9 10 85	A	(4,7)		
0043	0020	COLOR	0147	01	PARTITION COLOR CHIPS	1 10 85	1 10 85	A FYI	(5,7)		
0043	0020	GUIDE	0147B	02	COLOR GUIDE	8 22 85	9 10 85	A	(4,7)		
0043	0021	LETTER 2/20	0098C	02	MATERIAL DISTRIBUTORS	3 13 85	3 25 85	A FYI	(5)		
0043	0023	DATA	0918	01	VANITY C.B. BATHROOM	3 11 87	3 26 87	AAN-R	(3,7)		
0043	0023	DATA	0918A	02	VANITY C.B. BATHROOM	3 31 87	4 13 87	A	(4,7)		

Figure 8-10

200 Jackson Street
Camden, New Jersey 08104
Telephone 609/541-8100

PLANS TRANSMITTAL FORM

PAUL A. LAURENCE COMPANY
P.O. BOX 1267
10,000 HIGHWAY 55 WEST
MINNEAPOLIS, MINNESOTA 55440

Transmittal No. 827A

DATE October 9, 1986

TO OKA

JOB DELAWARE NO. 1 WPCF

LOCATION CAMDEN, N.J.

ORDER NO. CONTRACT 160

ATTN.: TOM STANGO

PALCO Job No. 9,66

We are sending by: Enclosed X MESSENGER ____ Separate Cover ____

For:
____ Approval
X Resubmittal
____ For your Files
____ As you Requested
____ Information Only
____ For Field Use
____ Please Return

____ Approved
____ Approved as Noted
____ Not Approved
____ Make Corrections
____ Resubmit for Final Approval
____ Resubmit ____ Copies
____ Send ____ Copies for Files, Field Use
____ Proceed with Fabrication
____ For Quotation
____ Measurements

Drawings as follows: (51.29) D-1793

NO. PRINTS	DWG NO	REV.NO	DESCRIPTION
7	84618A sheet 1 of 2		ICV-Local Control Panel. OKA I.D.#=51-0006
7	84618A sheet 2 of 2		RCV-Local Control Panel. OKA I.D.#=51-0007
7	84618B sheets 2 thru 4		WSV, RSV, & RIV-Local Control Panels
			OKA I.D.#=51-0009 thru 51-0011
7	COPIES		TECH. DATA

RECEIVED
FIELD OFFICE
1986 OCT -9 PM 1:
O'BRIEN-KREITZBERG

Remarks:

Copy of Transmittal to ____

Copy of Prints to ____

PAUL A. LAURENCE CO.

BY Marsh Funsho

Figure 8-11

including the specification section, the shop drawing, the transmittal number, the description, and the transmit-to-engineer date.

In order to correctly assign identification and submission numbers, the information taken from the contractor's transmittal form must be entered into a field log entitled "Shop Drawing Log Number Assignments" (Figure 8-12). Each shop drawing is listed in order and assigned a corresponding identification number. The appropriate submission number can also be determined. It is best to copy the information straight from the contractor's submittal form to guard against any claims by the contractor that the information is not correct. Next, the shop drawings themselves are stamped with the newly assigned computer identification number (Figure 8-13). All of the information supplied by the contractor and assigned by the owner's representative is entered onto computer input sheets (Figure 8-14).

The shop drawing can now be forwarded to the engineer. Attention must be paid to a quick turnaround to avoid any claims by the contractor and to keep the list current. The two remaining pieces of information, the transmit-from-engineer date and the action (Figure 8-15), are added to the record upon receipt of the engineer's return transmittal. Action codes are defined, in this case, on the engineer's transmittal and at the beginning of the computer run. Each drawing will also be stamped by the engineer to reflect the action. Resubmittals are treated the same way. Information is recorded from the contractor's submittal, the resubmittal shop drawing is given the same computer identification number as the original submittal, and it is assigned the next submittal number (Figure 8-16). The result is a computerized list arranged to show a complete history of each shop drawing submitted.

The greatest advantage of computerizing the shop drawing log is that the information may be manipulated in a number of ways. First, it facilitates the calculation of the time that drawings remain outstanding (Figure 8-17). Second, by using different sorts, a variety of reports can be made. A criticality report (Figure 8-18) is created by sorting on the calculated outstanding days. The contractor may also request a sort based on the transmittal number (Figure 8-19). The whole file can even be divided into one that reflects only outstanding drawings and their histories, and one that reflects only approved shop drawings with their respective histories (Figure 8-20).

CCMUA CONTRACT NO. ____________
SPEC. SECT. NO. ____________
PAGE NO. ____________

SHOP DRAWING LOG NUMBER ASSIGNMENTS

OKA NO.	P.A.L. NO.	DATE REC'D	DESCRIPTION	1ST SUBMISSION	2ND SUBMISSION	3RD SUBMISSION	4TH SUBMISSION
0001	010	10/2/84	ENCASEMENT REBAR DRAWINGS (SM-1) 96"	10/2/84	11/6/84 10	12/18/84	3/8/85 10C
				10/23/84 3	11/19/84 3	2/11/85 4	DISTRIBUTION
0002	015	10/12/84	INFLUENT & BYPASS CONDUIT PIPE FOUNDATIONS AT-1 (2-13 15-20)	10/12/84	11/6/84 15	2/22/88 15B	
				10/24/84 3	11/12/84 4	DISTRIBUTION	
0003	020	10/22/84	AT-2 REBAR JUNCTION CHAMBER (POUR 1)	10/22/84	1/4/85 20B	1/17/85 20D	2/13/85 20F
				10/31/84 3	1/16/85 6	1/30/85 3	2/19/85 A
0004	020	10/22/84	AT-3 REBAR JUNCTION CHAMBER (14)	10/22/84	11/28/84 20A	1/11/85 20C	2/6/85 20E
				10/31/84 3	12/12/84 3	1/15/85 4	DISTRIBUTION
0005	024	10/26/84	AT-4 REBAR BASE SLABS/DOWELS EFFL TO NORTH SPLITTER BOX (25)	10/26/84	2/28/84 24A		
				11/5/84 3	3/6/84 4		
0006	026	10/31/84	CB-1 COMPRESSOR BUILDING-PILE CAPS GRADE BEAMS/COLUMNS DOWELS (21-24)	10/31/84	12/17/84 26A	1/22/85 26C	
				11/5/84 3	1/8/85 4	DISTRIBUTION	
0007	026	10/31/84	CB-2 COMPRESSOR BUILDING-LABEL LIST/BENDING DETAILS	10/31/84	12/17/84 86A	1/21/85 26B	2/22/85 26D
				11/5/84 3	1/8/85 3	2/1/85 4	DISTRIBUTION
0008	027	10/31/84	CB-3 COMPRESSOR BUILDING COLUMNS/ COMPRESSOR SLAB (21-24)	10/31/84	12/17/84 27A	1/22/85 27B	
				11/5/84 3	1/8/85 4	DISTRIBUTION	
0009	028	10/31/84	AT-5 REBAR WALLS UP TO +15.15 @ EFFLUENT TO N. SPLITTER BOX (25)	10/31/84	12/17/84 28A	1/22/85 28B	
				11/6/84 2	1/8/85 4	DISTRIBUTION	

Figure 8-12

O'BRIEN-KREITZBERG
& ASSOC., INC.
CCMUA CONTRACT NO. 160
COMPUTER I.D. NO. S1-0009
ALL CORRESPONDENCE SHOULD
REFERENCE COMPUTER I.D. NO.

Figure 8-13

CCMUA SHOPFILE INPUT

160	Contract No.
0011	Spec Section
0006	Computer I.D. No.
CB-1	Drawing No.
0026	Part 1 Trans No.
	Part 2 Trans No.
01	Submittal No.
COMPRSSR BLDG P CAPS ETC	Description
1031 84	To Date
	From Date
	Action No. 1
	Action No. 2

160	Contract No.
0011	Spec Section
0007	Computer I.D. No.
CB-2	Drawing No.
0026	Part 1 Trans No.
	Part 2 Trans No.
01	Submittal No.
COMPRSSR BLDG LABEL LIST	Description
1031 84	To Date
	From Date
	Action No. 1
	Action No. 2

	Contract No.
	Spec Section
	Computer I.D. No.
	Drawing No.
	Part 1 Trans No.
	Part 2 Trans No.
	Submittal No.
	Description
	To Date
	From Date
	Action No. 1
	Action No. 2

Figure 8-14

Greeley and Hansen, Engineers
1818 Market Street, Suite 1360
Philadelphia, PA 19103
215/563-3460

G & P Transmittal No.
CCMUA-160-1765
PALCO No. 827A

TO: O'Brien-Kreitzberg & Assoc., Inc.
1502 Route 38
Cherry Hill, New Jersey 08002

Attn: Mr. Wesley F. Mikes

DATE: November 3, 1986

PROJECT: CCMUA Delaware No.1 WPCF Improvements
CONTRACT: 160-38 MGD

SECTION: W-80 – Instrumentation and Controls
S-1.29 – System Control Schemes

Gentlemen:

We are sending you herewith the following sheets:

Qty.	Drawing Number	OKA Computer I.D. Number	Title	Action Taken
5	84618A Sht 1	S1-0006	ICV – Local Panels	4,8
5	84618A Sht 2	S1-0007	RCV – Local Panels	4,8
5	84618B Sht 2	S1-0009	WSV, RSV & RIV – Local Panels	4,8
5	84618B Sht 3	S1-0010	WSV, RSV & RIV – Local Panels	4,8
5	84618B Sht 4	S1-0011	WSV, RSV & RIV – Local Panels	4,8
5		S1-0005	Instrumentation System Panels	
			84618A: BM01, 02	3,7
			84618B: BMO1	3,7
			All remaining sheets	4,8

Prepared by: PALCO/Essex Engineering

These are:
1. To be checked.
2. Examined and returned for correction.
3. Approved subject to corrections marked.
4. Approved.
5. For your information and use.
6. Incomplete submittal.

7. & 8. See remarks.

REMARKS:

7. Revise and resubmit corrected copies for final approval.
8. Approved for distribution.

cc: H. Engelbert – trans. only
A. Cevallos – trans. only
R. Soltysiak – trans. only

Yours very truly,
GREELEY AND HANSEN

By____________________

Figure 8-15

REPORT NO 1
LIST BY SHOP DWG —WITHOUT ERRORS

CAMDEN COUNTY MUNICIPAL UTILITIES AUTHORITY
PREPARED BY: O'BRIEN–KREITZBERG & ASSOC., INC.

PAGE 30
DATA DATE

PROJECT: DELAWARE NO. 1 WPCF CONTRACT: 160 CONTRACTOR: PAUL A. LAURENCE ENGINEER GREELEY & HANSEN

SPEC SECT	ID NO	SHOP DRAWING	TRANS NO.	SUB NO.	DESCRIPTION	TRANSMIT TO ENG	TRANSMIT FROM ENG	ACTION		OUTSTANDING CNTR	SHOPDRAWING ENG
S-1	0001	CCMUA-TEMP-E1	0219	01	TEMPORARY POWER PLAN	2 11 85	2 18 85	AAN-R	(3,7)		
S-1	0001	CCMUA-TEMP-E1	0219A	02	TEMPORARY POWER PLAN	2 28 85	3 11 85	A	(4,7)		
S-1	0002	CATALOG CUT	0219A	01	TEMP POWER METER DETAILS	2 28 85	3 11 85	A	(4,7)		
S-1	0003	CODE	0273	01	VALVE I.D. CODE	2 28 85	3 07 85	A	(4,7)		
S-1	0005	TECH DATA	0827	01	INSTRUMENTATION SYS PANEL	8 04 86	9 23 86	AAN-R	(3,7)		
S-1	0005	TECH DATA	0827A	02	INSTRUMENTATION SYS PANEL	10 09 86	11 03 86	AAN-R	(3,7)		
S-1	0005	TECH DATA	0827B	03	INSTRUMENTATION SYS PANEL	1 05 87	2 10 87	A	(4,7)		
S-1	0006	84618A SHT-1	0827	01	ICV LOCAL PANELS	8 04 86	9 23 86	AAN-R	(3,7)		
S-1	0006	84618A SHT-1	0827A	02	ICV LOCAL PANELS	10 09 86	11 03 86	A	(4,8)		
S-1	0007	84618A SHT-2	0827	01	ICV LOCAL PANELS	8 04 86	9 23 86	AAN-R	(3,7)		
S-1	0007	84618A SHT-2	0827A	02	RCV LOCAL PANELS	10 09 86	11 03 86	A	(4,8)		
S-1	0008	84618B SHT-1	0827	01	WIV LOCAL CNTRL PANELS	8 04 86	9 23 86	A	(4,8)		
S-1	0009	84618B SHT-2	0827	01	WIV LOCAL CNTRL PANELS	8 04 86	9 23 86	AAN-R	(3,7)		
S-1	0009	84618B	0827A	02	WSV, RSV, RIV-LOCAL CON PNL	10 09 86	11 03 86	A	(4,8)		
S-1	0010	84618B SHT-3	0827	01	WIV LOCAL CNTRL PANELS	8 04 86	9 23 86	AAN-R	(3,7)		
S-1	0010	84618B	0827A	02	WSV, RSV, RIV-LOCAL CON PNL	10 09 86	11 03 86	A	(4,8)		
S-1	0011	84618B SHT-4	0827	01	WIV LOCAL CNTRL PANELS	8 04 86	9 23 86	AAN-R	(3,7)		
S-1	0011	84618B	0827A	02	WSV, RSV, RIV-LOCAL CON PNL	10 09 86	11 03 86	A	(4,8)		
S-1	0012	SAMPLE	0896	01	FIBERGLASS ID NAMEPLATES	1 30 87	2 10 87	R & R	(7)		
S-1	0012	SAMPLE	0896A	02	FIBERGLASS ID NAMEPLATES	6 04 87	5 22 87	A	(4,7)		
S-1	0013	SCHEDULE	0917	01	PALCO VALUE I.D. SCHEDULE	3 11 87	5 29 87	R & R	(2,7)		
S-1	0013		0917A	02	PALCO VALVE ID SCHEDULE	6 16 87	7 16 87	A	(4,7)		
S-1	0013	SCHEDULE	0917B	03	PALCO VALUE I.D. SCHEDULE	8 21 87	10 28 87	A	(4,9)		
S-1	0014	SCHEDULE	0917	01	WILLARD VALUE I.D. SCH	3 11 87	5 29 87	R & R	(2,7)		
S-1	0014	SCHEDULE	0917B	02	WILLARD I.D. SCHEDULE	8 21 87	11 20 87	AAN-R	(3,7)		
S-1	0015		0829A	01	EQUIP SCH. FOR TAGGING	8 19 87	9 15 87	A	(7)	45	
S-1	0015	SCHEDULE	0829B	02	EQUIP SCH FOR TAGGING	11 11 87	11 30 87	A	(7)		
S-1	0016	SCHEDULE	0917B	01	VALVE ID SCHEDULE OXYGEN	9 21 87	10 28 87	AAN-R	(3,7)		
S-1	0016	SCHEDULE	0917C	02	VALVE ID SCHEDULE OXYGEN	11 11 87	11 18 87	A	(4,7)		

Figure 8-16

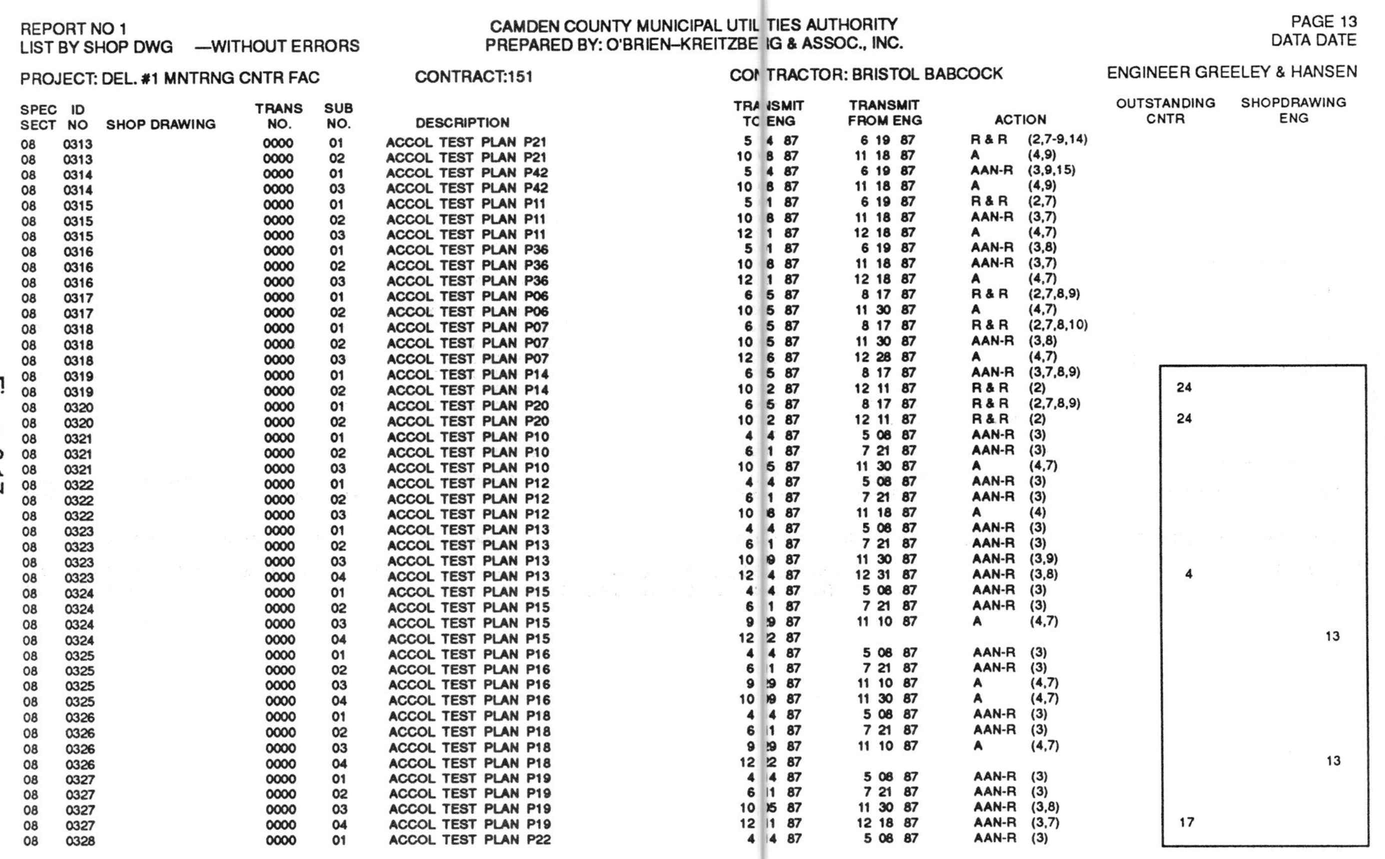

REPORT NO 1
LIST BY SHOP DWG —WITHOUT ERRORS

CAMDEN COUNTY MUNICIPAL UTILITIES AUTHORITY
PREPARED BY: O'BRIEN–KREITZBERG & ASSOC., INC.

PAGE 13
DATA DATE

PROJECT: DEL. #1 MNTRNG CNTR FAC   CONTRACT:151   CONTRACTOR: BRISTOL BABCOCK   ENGINEER GREELEY & HANSEN

SPEC SECT	ID NO	SHOP DRAWING	TRANS NO.	SUB NO.	DESCRIPTION	TRANSMIT TO ENG	TRANSMIT FROM ENG	ACTION		OUTSTANDING CNTR	SHOPDRAWING ENG
08	0313		0000	01	ACCOL TEST PLAN P21	5 4 87	6 19 87	R & R	(2,7-9,14)		
08	0313		0000	02	ACCOL TEST PLAN P21	10 8 87	11 18 87	A	(4,9)		
08	0314		0000	01	ACCOL TEST PLAN P42	5 4 87	6 19 87	AAN-R	(3,9,15)		
08	0314		0000	03	ACCOL TEST PLAN P42	10 8 87	11 18 87	A	(4,9)		
08	0315		0000	01	ACCOL TEST PLAN P11	5 1 87	6 19 87	R & R	(2,7)		
08	0315		0000	02	ACCOL TEST PLAN P11	10 8 87	11 18 87	AAN-R	(3,7)		
08	0315		0000	03	ACCOL TEST PLAN P11	12 1 87	12 18 87	A	(4,7)		
08	0316		0000	01	ACCOL TEST PLAN P36	5 1 87	6 19 87	AAN-R	(3,8)		
08	0316		0000	02	ACCOL TEST PLAN P36	10 8 87	11 18 87	AAN-R	(3,7)		
08	0316		0000	03	ACCOL TEST PLAN P36	12 1 87	12 18 87	A	(4,7)		
08	0317		0000	01	ACCOL TEST PLAN P06	6 5 87	8 17 87	R & R	(2,7,8,9)		
08	0317		0000	02	ACCOL TEST PLAN P06	10 5 87	11 30 87	A	(4,7)		
08	0318		0000	01	ACCOL TEST PLAN P07	6 5 87	8 17 87	R & R	(2,7,8,10)		
08	0318		0000	02	ACCOL TEST PLAN P07	10 5 87	11 30 87	AAN-R	(3,8)		
08	0318		0000	03	ACCOL TEST PLAN P07	12 6 87	12 28 87	A	(4,7)		
08	0319		0000	01	ACCOL TEST PLAN P14	6 5 87	8 17 87	AAN-R	(3,7,8,9)		
08	0319		0000	02	ACCOL TEST PLAN P14	10 2 87	12 11 87	R & R	(2)	24	
08	0320		0000	01	ACCOL TEST PLAN P20	6 5 87	8 17 87	R & R	(2,7,8,9)		
08	0320		0000	02	ACCOL TEST PLAN P20	10 2 87	12 11 87	R & R	(2)	24	
08	0321		0000	01	ACCOL TEST PLAN P10	4 4 87	5 08 87	AAN-R	(3)		
08	0321		0000	02	ACCOL TEST PLAN P10	6 1 87	7 21 87	AAN-R	(3)		
08	0321		0000	03	ACCOL TEST PLAN P10	10 5 87	11 30 87	A	(4,7)		
08	0322		0000	01	ACCOL TEST PLAN P12	4 4 87	5 08 87	AAN-R	(3)		
08	0322		0000	02	ACCOL TEST PLAN P12	6 1 87	7 21 87	AAN-R	(3)		
08	0322		0000	03	ACCOL TEST PLAN P12	10 8 87	11 18 87	A	(4)		
08	0323		0000	01	ACCOL TEST PLAN P13	4 4 87	5 08 87	AAN-R	(3)		
08	0323		0000	02	ACCOL TEST PLAN P13	6 1 87	7 21 87	AAN-R	(3)		
08	0323		0000	03	ACCOL TEST PLAN P13	10 9 87	11 30 87	AAN-R	(3,9)		
08	0323		0000	04	ACCOL TEST PLAN P13	12 4 87	12 31 87	AAN-R	(3,8)	4	
08	0324		0000	01	ACCOL TEST PLAN P15	4 4 87	5 08 87	AAN-R	(3)		
08	0324		0000	02	ACCOL TEST PLAN P15	6 1 87	7 21 87	AAN-R	(3)		
08	0324		0000	03	ACCOL TEST PLAN P15	9 9 87	11 10 87	A	(4,7)		
08	0324		0000	04	ACCOL TEST PLAN P15	12 2 87					13
08	0325		0000	01	ACCOL TEST PLAN P16	4 4 87	5 08 87	AAN-R	(3)		
08	0325		0000	02	ACCOL TEST PLAN P16	6 1 87	7 21 87	AAN-R	(3)		
08	0325		0000	03	ACCOL TEST PLAN P16	9 9 87	11 10 87	A	(4,7)		
08	0325		0000	04	ACCOL TEST PLAN P16	10 9 87	11 30 87	A	(4,7)		
08	0326		0000	01	ACCOL TEST PLAN P18	4 4 87	5 08 87	AAN-R	(3)		
08	0326		0000	02	ACCOL TEST PLAN P18	6 1 87	7 21 87	AAN-R	(3)		
08	0326		0000	03	ACCOL TEST PLAN P18	9 9 87	11 10 87	A	(4,7)		
08	0326		0000	04	ACCOL TEST PLAN P18	12 2 87					13
08	0327		0000	01	ACCOL TEST PLAN P19	4 4 87	5 08 87	AAN-R	(3)		
08	0327		0000	02	ACCOL TEST PLAN P19	6 1 87	7 21 87	AAN-R	(3)		
08	0327		0000	03	ACCOL TEST PLAN P19	10 5 87	11 30 87	AAN-R	(3,8)		
08	0327		0000	04	ACCOL TEST PLAN P19	12 1 87	12 18 87	AAN-R	(3,7)	17	
08	0328		0000	01	ACCOL TEST PLAN P22	4 4 87	5 08 87	AAN-R	(3)		

Figure 8-17

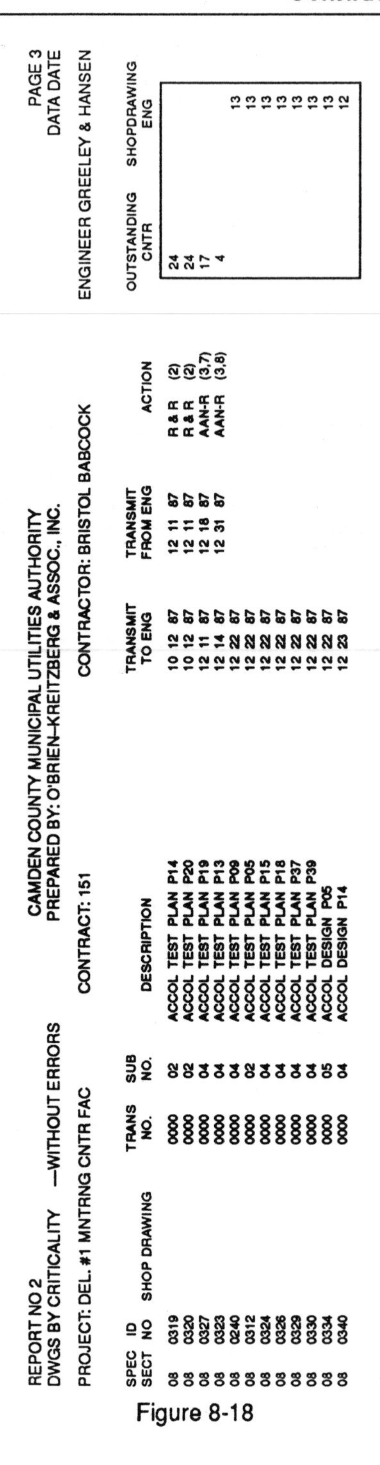

REPORT NO 2 — CAMDEN COUNTY MUNICIPAL UTILITIES AUTHORITY — PAGE 3
DWGS BY CRITICALITY —WITHOUT ERRORS — PREPARED BY: O'BRIEN–KREITZBERG & ASSOC., INC. — DATA DATE

PROJECT: DEL. #1 MNTRNG CNTR FAC — CONTRACT: 151 — CONTRACTOR: BRISTOL BABCOCK — ENGINEER GREELEY & HANSEN

SPEC SECT	ID NO	SHOP DRAWING	TRANS NO.	SUB NO.	DESCRIPTION	TRANSMIT TO ENG	TRANSMIT FROM ENG	ACTION	OUTSTANDING CNTR	SHOPDRAWING ENG
08	0319		0000	02	ACCOL TEST PLAN P14	10 12 87	12 11 87	R & R (2)	24	
08	0320		0000	02	ACCOL TEST PLAN P20	10 12 87	12 11 87	R & R (2)	24	
08	0327		0000	04	ACCOL TEST PLAN P19	12 11 87	12 18 87	AAN-R (3,7)	17	
08	0323		0000	04	ACCOL TEST PLAN P13	12 14 87	12 31 87	AAN-R (3,8)	4	
08	0240		0000	04	ACCOL TEST PLAN P09	12 22 87				13
08	0312		0000	02	ACCOL TEST PLAN P05	12 22 87				13
08	0324		0000	04	ACCOL TEST PLAN P15	12 22 87				13
08	0326		0000	04	ACCOL TEST PLAN P18	12 22 87				13
08	0329		0000	04	ACCOL TEST PLAN P37	12 22 87				13
08	0330		0000	04	ACCOL TEST PLAN P39	12 22 87				13
08	0334		0000	05	ACCOL DESIGN P05	12 22 87				13
08	0340		0000	04	ACCOL DESIGN P14	12 23 87				12

Figure 8-18

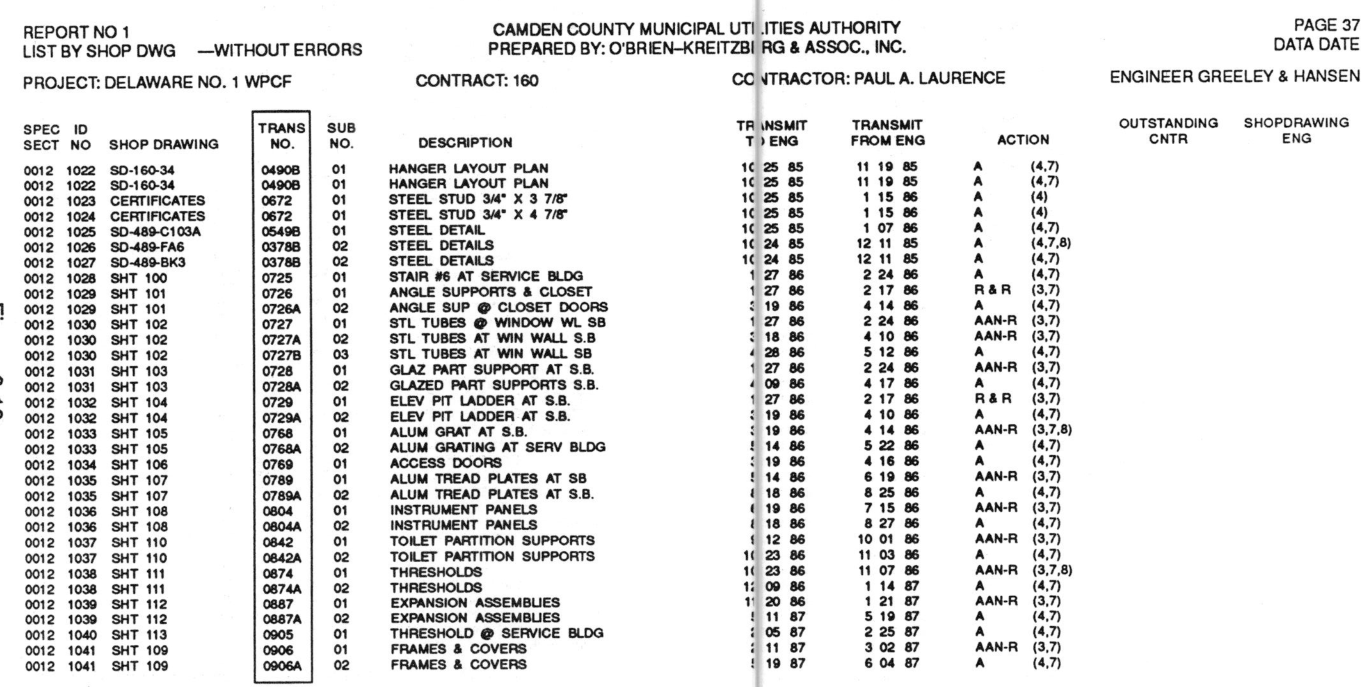

REPORT NO 1
LIST BY SHOP DWG —WITHOUT ERRORS

CAMDEN COUNTY MUNICIPAL UTI[illegible]ITIES AUTHORITY
PREPARED BY: O'BRIEN–KREITZBI[illegible]RG & ASSOC., INC.

PAGE 37
DATA DATE

PROJECT: DELAWARE NO. 1 WPCF CONTRACT: 160 CO[illegible]NTRACTOR: PAUL A. LAURENCE ENGINEER GREELEY & HANSEN

SPEC SECT	ID NO	SHOP DRAWING	TRANS NO.	SUB NO.	DESCRIPTION	TR[illegible]NSMIT T[illegible] ENG	TRANSMIT FROM ENG	ACTION		OUTSTANDING CNTR	SHOPDRAWING ENG
0012	1022	SD-160-34	0490B	01	HANGER LAYOUT PLAN	1[illegible] 25 85	11 19 85	A	(4,7)		
0012	1022	SD-160-34	0490B	01	HANGER LAYOUT PLAN	1[illegible] 25 85	11 19 85	A	(4,7)		
0012	1023	CERTIFICATES	0672	01	STEEL STUD 3/4" X 3 7/8"	1[illegible] 25 85	1 15 86	A	(4)		
0012	1024	CERTIFICATES	0672	01	STEEL STUD 3/4" X 4 7/8"	1[illegible] 25 85	1 15 86	A	(4)		
0012	1025	SD-489-C103A	0549B	01	STEEL DETAIL	1[illegible] 25 85	1 07 86	A	(4,7)		
0012	1026	SD-489-FA6	0378B	02	STEEL DETAILS	1[illegible] 24 85	12 11 85	A	(4,7,8)		
0012	1027	SD-489-BK3	0378B	02	STEEL DETAILS	1[illegible] 24 85	12 11 85	A	(4,7)		
0012	1028	SHT 100	0725	01	STAIR #6 AT SERVICE BLDG	1 27 86	2 24 86	A	(4,7)		
0012	1029	SHT 101	0726	01	ANGLE SUPPORTS & CLOSET	1 27 86	2 17 86	R & R	(3,7)		
0012	1029	SHT 101	0726A	02	ANGLE SUP @ CLOSET DOORS	[illegible] 19 86	4 14 86	A	(4,7)		
0012	1030	SHT 102	0727	01	STL TUBES @ WINDOW WL SB	1 27 86	2 24 86	AAN-R	(3,7)		
0012	1030	SHT 102	0727A	02	STL TUBES AT WIN WALL S.B	[illegible] 18 86	4 10 86	AAN-R	(3,7)		
0012	1030	SHT 102	0727B	03	STL TUBES AT WIN WALL SB	[illegible] 28 86	5 12 86	A	(4,7)		
0012	1031	SHT 103	0728	01	GLAZ PART SUPPORT AT S.B.	1 27 86	2 24 86	AAN-R	(3,7)		
0012	1031	SHT 103	0728A	02	GLAZED PART SUPPORTS S.B.	[illegible] 09 86	4 17 86	A	(4,7)		
0012	1032	SHT 104	0729	01	ELEV PIT LADDER AT S.B.	1 27 86	2 17 86	R & R	(3,7)		
0012	1032	SHT 104	0729A	02	ELEV PIT LADDER AT S.B.	[illegible] 19 86	4 10 86	A	(4,7)		
0012	1033	SHT 105	0768	01	ALUM GRAT AT S.B.	[illegible] 19 86	4 14 86	AAN-R	(3,7,8)		
0012	1033	SHT 105	0768A	02	ALUM GRATING AT SERV BLDG	[illegible] 14 86	5 22 86	A	(4,7)		
0012	1034	SHT 106	0769	01	ACCESS DOORS	[illegible] 19 86	4 16 86	A	(4,7)		
0012	1035	SHT 107	0789	01	ALUM TREAD PLATES AT SB	[illegible] 14 86	6 19 86	AAN-R	(3,7)		
0012	1035	SHT 107	0789A	02	ALUM TREAD PLATES AT S.B.	[illegible] 18 86	8 25 86	A	(4,7)		
0012	1036	SHT 108	0804	01	INSTRUMENT PANELS	[illegible] 19 86	7 15 86	AAN-R	(3,7)		
0012	1036	SHT 108	0804A	02	INSTRUMENT PANELS	[illegible] 18 86	8 27 86	A	(4,7)		
0012	1037	SHT 110	0842	01	TOILET PARTITION SUPPORTS	[illegible] 12 86	10 01 86	AAN-R	(3,7)		
0012	1037	SHT 110	0842A	02	TOILET PARTITION SUPPORTS	1[illegible] 23 86	11 03 86	A	(4,7)		
0012	1038	SHT 111	0874	01	THRESHOLDS	1[illegible] 23 86	11 07 86	AAN-R	(3,7,8)		
0012	1038	SHT 111	0874A	02	THRESHOLDS	1[illegible] 09 86	1 14 87	A	(4,7)		
0012	1039	SHT 112	0887	01	EXPANSION ASSEMBLIES	1[illegible] 20 86	1 21 87	AAN-R	(3,7)		
0012	1039	SHT 112	0887A	02	EXPANSION ASSEMBLIES	[illegible] 11 87	5 19 87	A	(4,7)		
0012	1040	SHT 113	0905	01	THRESHOLD @ SERVICE BLDG	[illegible] 05 87	2 25 87	A	(4,7)		
0012	1041	SHT 109	0906	01	FRAMES & COVERS	[illegible] 11 87	3 02 87	AAN-R	(3,7)		
0012	1041	SHT 109	0906A	02	FRAMES & COVERS	[illegible] 19 87	6 04 87	A	(4,7)		

Figure 8-19

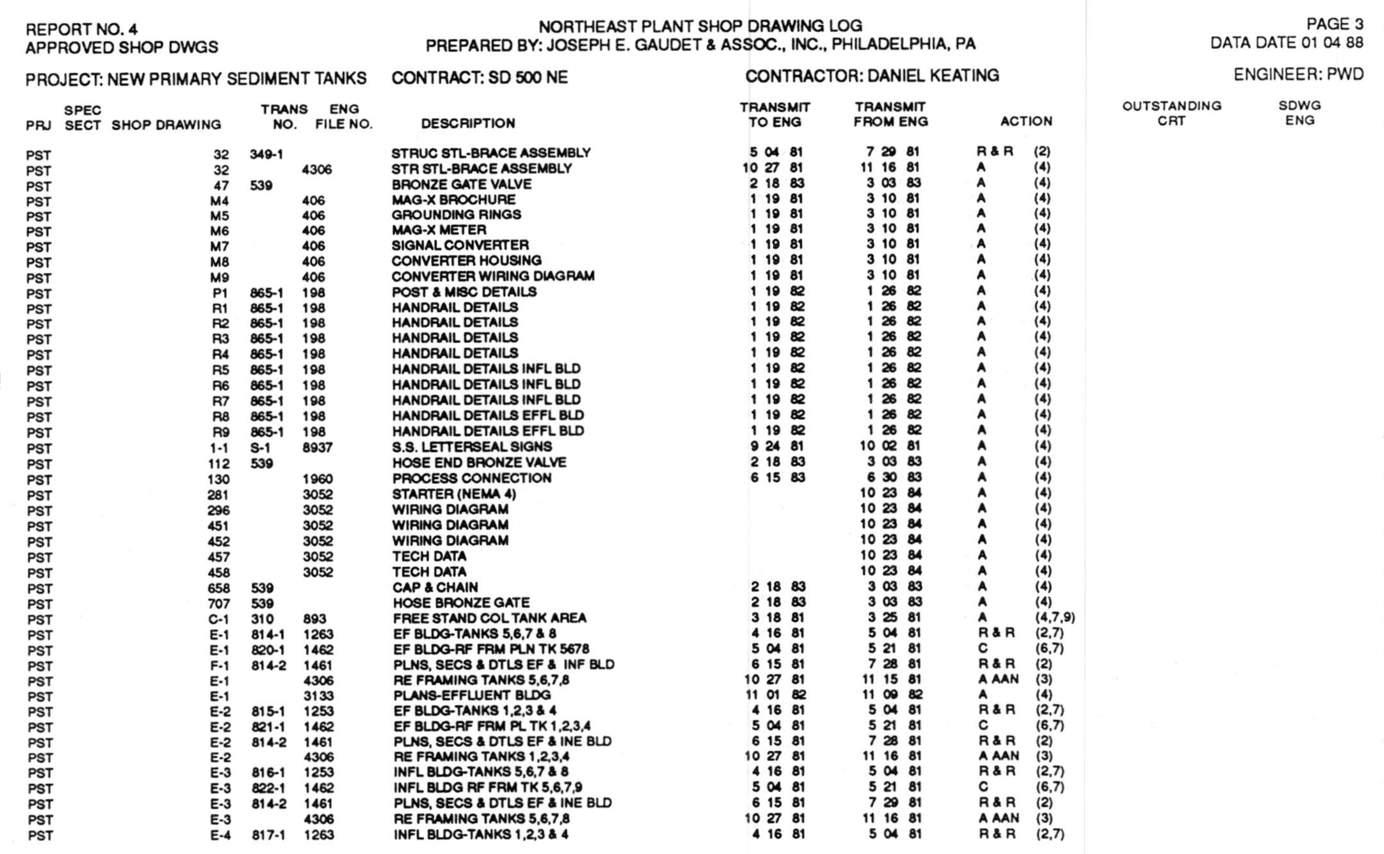

REPORT NO. 4
APPROVED SHOP DWGS

NORTHEAST PLANT SHOP DRAWING LOG
PREPARED BY: JOSEPH E. GAUDET & ASSOC., INC., PHILADELPHIA, PA

PAGE 3
DATA DATE 01 04 88

PROJECT: NEW PRIMARY SEDIMENT TANKS   CONTRACT: SD 500 NE   CONTRACTOR: DANIEL KEATING   ENGINEER: PWD

PRJ	SPEC SECT	SHOP DRAWING	TRANS NO.	ENG FILE NO.	DESCRIPTION	TRANSMIT TO ENG	TRANSMIT FROM ENG	ACTION		OUTSTANDING CRT	SDWG ENG
PST		32	349-1		STRUC STL-BRACE ASSEMBLY	5 04 81	7 29 81	R & R	(2)		
PST		32		4306	STR STL-BRACE ASSEMBLY	10 27 81	11 16 81	A	(4)		
PST		47	539		BRONZE GATE VALVE	2 18 83	3 03 83	A	(4)		
PST		M4		406	MAG-X BROCHURE	1 19 81	3 10 81	A	(4)		
PST		M5		406	GROUNDING RINGS	1 19 81	3 10 81	A	(4)		
PST		M6		406	MAG-X METER	1 19 81	3 10 81	A	(4)		
PST		M7		406	SIGNAL CONVERTER	1 19 81	3 10 81	A	(4)		
PST		M8		406	CONVERTER HOUSING	1 19 81	3 10 81	A	(4)		
PST		M9		406	CONVERTER WIRING DIAGRAM	1 19 81	3 10 81	A	(4)		
PST		P1	865-1	198	POST & MISC DETAILS	1 19 82	1 26 82	A	(4)		
PST		R1	865-1	198	HANDRAIL DETAILS	1 19 82	1 26 82	A	(4)		
PST		R2	865-1	198	HANDRAIL DETAILS	1 19 82	1 26 82	A	(4)		
PST		R3	865-1	198	HANDRAIL DETAILS	1 19 82	1 26 82	A	(4)		
PST		R4	865-1	198	HANDRAIL DETAILS	1 19 82	1 26 82	A	(4)		
PST		R5	865-1	198	HANDRAIL DETAILS INFL BLD	1 19 82	1 26 82	A	(4)		
PST		R6	865-1	198	HANDRAIL DETAILS INFL BLD	1 19 82	1 26 82	A	(4)		
PST		R7	865-1	198	HANDRAIL DETAILS INFL BLD	1 19 82	1 26 82	A	(4)		
PST		R8	865-1	198	HANDRAIL DETAILS EFFL BLD	1 19 82	1 26 82	A	(4)		
PST		R9	865-1	198	HANDRAIL DETAILS EFFL BLD	1 19 82	1 26 82	A	(4)		
PST		1-1	S-1	8937	S.S. LETTERSEAL SIGNS	9 24 81	10 02 81	A	(4)		
PST		112	539		HOSE END BRONZE VALVE	2 18 83	3 03 83	A	(4)		
PST		130		1960	PROCESS CONNECTION	6 15 83	6 30 83	A	(4)		
PST		281		3052	STARTER (NEMA 4)		10 23 84	A	(4)		
PST		296		3052	WIRING DIAGRAM		10 23 84	A	(4)		
PST		451		3052	WIRING DIAGRAM		10 23 84	A	(4)		
PST		452		3052	WIRING DIAGRAM		10 23 84	A	(4)		
PST		457		3052	TECH DATA		10 23 84	A	(4)		
PST		458		3052	TECH DATA		10 23 84	A	(4)		
PST		658	539		CAP & CHAIN	2 18 83	3 03 83	A	(4)		
PST		707	539		HOSE BRONZE GATE	2 18 83	3 03 83	A	(4)		
PST		C-1	310	893	FREE STAND COL TANK AREA	3 18 81	3 25 81	A	(4,7,9)		
PST		E-1	814-1	1263	EF BLDG-TANKS 5,6,7 & 8	4 16 81	5 04 81	R & R	(2,7)		
PST		E-1	820-1	1462	EF BLDG-RF FRM PLN TK 5678	5 04 81	5 21 81	C	(6,7)		
PST		F-1	814-2	1461	PLNS, SECS & DTLS EF & INF BLD	6 15 81	7 28 81	R & R	(2)		
PST		E-1		4306	RE FRAMING TANKS 5,6,7,8	10 27 81	11 15 81	A AAN	(3)		
PST		E-1		3133	PLANS-EFFLUENT BLDG	11 01 82	11 09 82	A	(4)		
PST		E-2	815-1	1253	EF BLDG-TANKS 1,2,3 & 4	4 16 81	5 04 81	R & R	(2,7)		
PST		E-2	821-1	1462	EF BLDG-RF FRM PL TK 1,2,3,4	5 04 81	5 21 81	C	(6,7)		
PST		E-2	814-2	1461	PLNS, SECS & DTLS EF & INE BLD	6 15 81	7 28 81	R & R	(2)		
PST		E-2		4306	RE FRAMING TANKS 1,2,3,4	10 27 81	11 16 81	A AAN	(3)		
PST		E-3	816-1	1253	INFL BLDG-TANKS 5,6,7 & 8	4 16 81	5 04 81	R & R	(2,7)		
PST		E-3	822-1	1462	INFL BLDG RF FRM TK 5,6,7,9	5 04 81	5 21 81	C	(6,7)		
PST		E-3	814-2	1461	PLNS, SECS & DTLS EF & INE BLD	6 15 81	7 29 81	R & R	(2)		
PST		E-3		4306	RE FRAMING TANKS 5,6,7,8	10 27 81	11 16 81	A AAN	(3)		
PST		E-4	817-1	1263	INFL BLDG-TANKS 1,2,3 & 4	4 16 81	5 04 81	R & R	(2,7)		

Figure 8-20

APPENDIX 8A

CONSTRUCTION PHASE SAMPLE PROGRESS MEETING MINUTES

Meeting Minutes 01/05/89

I. Subject — Progress Meeting No. 23

II. Date — 01/04/89

III. Time — 8:30 AM

IV. Place — Project Site

V. Attendees — See Attached List

VI. Notes:

RESPONSIBILITY	CATEGORY	ITEM
	A. SUBCONTRACTOR ISSUES	
	OLD BUSINESS — None	
	NEW BUSINESS	
WCC	23.641	Mechanical; RFI 279 Non-Pressure Balance Valve: The non-pressure balance valves were rejected by the Architect but RFI 279 requires installation. The Architect confirmed that RFI 279 should be followed and that the submittal should be resubmitted for approval.
WCC/DDP	23.642	Mechanical; Hose Bids: It was confirmed that the contract will use the original design for the hose bids. WCC will cut the block and grout. RFI #2072 needs to be reissued by the Architect with a detail.
WCC	23.643	RFI 63; Access Panels: Access panels in the sallyport area are required per the code according to the Architect due to the slip fittings and valves. Contractor states that may be a conflict with the pistol lockers.
	B. SUBMITTALS	
	OLD BUSINESS	
WCC	7.138	**Masonry Imbeds (05300):** Building No. 1 needed.
WCC	21.575	**Submittal 12A:** Boiler Tanks: Rejected by the Architect. WCC states items being requested are in excess of specification 15144. AE to review again.

WCC	18.428	**Submittal 62:** Tube: Waiting on information from WCC concerning tube configuration. AE is assisting in the information process.
WCC	18.429	**Submittal 111:** WCC to resubmit.
DDP	19.475	**Submittal 128:** The directional signage was due 11/14. Not critical according to WCC. Will be returned on 01/06.
DDP	19.477	**Submittal 133:** Elevator call button was due on 11/22. Fireman key switch was defined by RFI 1117. Will be returned on 01/06. Not critical according to WCC.
WCC	19.519	**Submittal 148A:** Carpet: Was resubmitted to AE but will be rejected.
DDP	20.529	**Submittal 153:** Cabinets: AE requested five weeks to review. WCC agreed. Submittal due 1/11.
WCC	21.576	**Building 1:** The Fire Marshall has partially reviewed the fire protection drawings. WCC holding for complete package.
WCC	22.603	The Architect requested a list of remaining submittals from WCC.
	NEW BUSINESS	
WCC DDP DDP	23.644	**Submittal Status:** 1) Answered: See Attachment A1 2) Received: See Attachment A2 3) Overdue: See Attachment A3
WCC	23.645	**Submittal 156:** Wood Doors were due on 12/29 and was returned on 01/04/89.
WCC	23.646	**Submittal 157:** Metal Deck was due on 12/30 and was returned on 01/04/89.
WCC	23.647	**Submittal 105:** Masonry Mock-Up: OK states that a copy is needed for the record.
	C. RFI's	
	OLD BUSINESS	
WCC	15.338	**Topsoil Amendments:** WCC to provide additional test data.
DDP/WCC	17.410	**RFI 114:** Cabinet hardware substitution: Was due on 09/30. Not critical according with WCC. AE requested samples of substitution for review prior to answering the RFI.

WCC	17.425	**Kidde Multiplex Submittal RFI 206:** WCC indicates a change will be requested.
WCC	21.567	**Masonry:** RFI #2071: 16 inch lintel blocks are required. 8" blocks installed to date can remain.
DDP	22.607	**RFI 47A:** Power files: Due on December 15. Will be returned on January 11th. No impact according to WCC. AE to meet with Gerry concerning led function.
DDP	22.612	**RFI 274:** Roof Hatch for Building 4 was due on December 16. Will be returned on January 3rd. No impact according to WCC.
WCC	22.613	**RFI 236A:** RFI 236A: Brinks 3020 Lock/Sergeant cylinder: AE to obtain comments from sergeant concerning compatibility.
	NEW BUSINESS	
WCC DDP DDP	23.648	**RFI STATUS:** 1) Answered: See Attachment B1 2) Received: See Attachment B2 3) Overdue: See Attachment B3
DDP	23.649	**RFI 39B:** 4" pipe vs. 4" stud: Was due on 12/21 and will be returned on 01/06. Not critical according to WCC.
DDP	23.650	**RFI 63B:** Access panel vs. CMU: Was due on 12/30 and will be returned on 01/06/89.
DDP	23.651	**RFI 234A:** Due on 01/04. Will be returned on 01/11/89. Not critical according to WCC.
DDP	23.652	**RFI's 256 and 257:** Were due on 12/20. Will be returned on 01/06. Not critical according to WCC.

D. PCO's

OLD BUSINESS

OK/WCC	22.599	**Mechanical:** Day Tank: Contractor indicates a 16 week lead time. PC being negotiated.
WCC	1.9	**VE Rail Change:** To be submitted.
WCC	1.33	**Labor Rates:** Partially submitted.
DDP/OK	14.303	**RFI 2044:** Seven windows: PC to be revised for aluminum and HM Frames. AE to revise drawings for frames.
WCC/DDP	17.399	**Kitchen RFI 1103:** Hold on cart equipment waiting for price on quick chiller. PC to be issued.

NEW BUSINESS

	23.653	**PC Status:** See Attachment C.
WCC		1) **Pricing needed:** 15, 17, 19, 25, 26, 27, 29, 30, 32, 33, 34, 37, 43, 44, 47, 36, 39, 48, 49, 51, 52, 53 and 55.
OK		2) **To Be Issued:** RFI's 1012 and 2044.
OK		3) **Change Orders to be Issued:** 1, 4, 11, 23, 31, 40, 45 and 46.
OK/WCC		4) **To Be Negotiated:** 2, 12, 13, 14, 21, 38 and 54.
		5) **To Be Reviewed:** 0
WCC		6) **Change Orders with WCC:** 22, 8, 11, 24, 35, 41 and 42.
DDP		7) **Changer Orders with AE:** 3, 6, 10, 16, 18 and 20.
		8) **RFI Rejected as Changed:** 0
DDP	23.654	The AE was requested to return all change orders within 5 working days. PC's 3, 6, 10, 16, 18 and 20 will be returned by 1/11.

E. **SCHEDULE**

OLD BUSINESS

WCC	18.466	WCC was requested to revise the CPM to reflect the major subcontractor durations and manpower. WCC did not submit by 01/04. Will submit ASAP.

NEW BUSINESS

WCC	22.617	**Schedule Status:** See Attachment D-1. WCC presented a two week schedule dated 01/03/89, which indicated the critical path and other areas which were behind schedule.
WCC	23.656	**Item AK Steel Windows:** Should not be on the critical path according to WCC.
WCC	23.657	**Item C FRP Footings:** 95% complete.
WCC	23.658	**Item D Conduit and Raise Floor:** Delayed by other slab construction.
WCC	23.659	**Item F CMU First Course:** 98% complete.
WCC	23.660	**Item G Underground Electrical:** 98% complete.

WCC 23.661 **Item J Underground Communication:** 98% complete.

WCC 23.662 **Item L FRP:** Pour No. 4 on 01/03, Pour No. 5 on 01/12. Generator room on 01/17.

WCC 23.663 **Item Y Electrical Communication First Course:** Substantial complete.

WCC 23.664 **OK Schedule Status Report:** See Attachment D-2. The OK Schedule Analysis indicates that the Contractor is approximately 30 calendar days behind schedule.

WCC 23.665 **Reasons for Delay:** Building 1 concrete Pour #4 has slipped from 12/12 to 01/03 due to rain and other factors. The Contractor's Pour No. 5 will slip from 01/05 to 01/12. The Contractor's 30 day calendar delay is due to the slippage of the pours as well as the preceding underslab electrical and communication installation. The contractor's stated that they will be planning to make up the delay with increased masonry production.

WCC 23.666 **Controlling Activities:** The Controlling Activities for the current week are the underslab electrical and communications which are holding up the slab work and the block work on the east side of Building No. 1.

WCC 23.667 **Manpower:** Again it was noted that electrical and communications manpower was understaffed and continue to delay the critical path progress.

F. SAFETY

OLD BUSINESS — None

NEW BUSINESS

WCC 23.668 The Contractor was requested to clean Giant Highway.

G. QUALITY CONTROL

OLD BUSINESS

WCC 22.614 **RFI 286 Door Position Switch:** The Contractor indicated that the doors were prepared for the wrong door position switch. The County stated that the Universal Security switch is not an equivalent and that Foldger Adams switch must be installed.

NEW BUSINESS — None

H. **PAYMENT**

OLD BUSINESS

WCC	14.328	**Letter of Credit:** Received from contractor and sent to County Council.

NEW BUSINESS

OK	23.669	A rough draft of the December payment was received.

I. **GENERAL**

OLD BUSINESS

WCC	1.3	**As-Builts of Project A:** WCC states drawings are not accurate.
WCC	16.367	**Electrical Equipment Code Spacing Claim:** WCC issued claim for seven days and notification of change for foundation removal. CM final decision has been issued.
WCC/OK	22.637	Street lights are not currently working and may have been cut.

NEW BUSINESS

OK	23.670	**Sales Tax:** County Council has been requested to issue an opinion concerning sales tax.
WCC	23.671	The CM trailor has 3 leaks.
OK	23.672	**Tree Selection:** A meeting will be set up by Gerry among the Architect, landscape consultant and landscape sub-contractor to select the trees. A 30 day notice received from WCC.
OK	23.673	A Telephone meeting will be held to start scheduling the telephone requirements.
DDP	23.674	**Cell Doors in Building No. 4:** The Architect indicated that the security consultant is suggesting that the strike plate be changed for all of the security doors in Building No. 4. AE will confirm and issue RFI if required.

Attachments: Attendees
Submittal Log (Attachments A1–A3)
RFI Log (Attachments B1–B4)
PCO Log (Attachment C)
WCC Two Week Schedule (Attachment D-1)
OK Schedule Status Report (Attachment D-2)
FI Log (Attachment E)
Exception Notices (Attachment F)

mtg/mm23.min

CHAPTER 9

Documentation for Acceptance of Work

WARRANTIES AND GUARANTEES

§ 9.1 One month prior to the forecast substantial completion date, the construction manager will write a letter notifying the contractor of the requirements for warranty/guarantee submittals under the contract. This letter should include a list of required warranties by which to monitor the contractor's compliance with the warranty submittal provisions.

The contractor will submit warranties/guarantees within the time specified by the contract and prior to issuance of the final payment. The field engineer should take this opportunity to assure that all letters of certification required either during the submittal or during the field installation processes have been provided by the contractor.

If during the warranty period (which is initiated with the date of substantial completion) any work is found to be defective or not in accordance with the contract requirements, the field engineer will notify the contractor in writing of the contractor's responsibility to perform corrective work in accordance with the contract provisions.

OPERATIONS AND MAINTENANCE MANUALS

§ 9.2 Contracts usually require the contractor to provide one copy of the operations and maintenance manuals at least 30 days in advance of substantial completion and/or training. This allows the construction manager field engineers and owner operating staff to become familiar with the system prior to final inspections and training, and provide comments for final submission.

Notwithstanding the construction contract requirements, the construction manager should, approximately 90 days before substantial completion or training, prepare a list of operations and maintenance manuals required by the specifications. This list should be organized to provide for ready identification of the documents and the applicable specification sections. The field engineer should provide the contractor with this information, citing the applicable specification section requiring submittal of the operations and maintenance manuals, and establishing a "due date" for the material. Upon receipt of these materials from the contractor, depending on the contract requirements, the field engineer should either review the submittal for format, content, and consistency with the specification, or review the submittal for specification consistency and transmit it to the architect/engineer to review the content.

PROJECT CLOSEOUT

§ 9.3 Project closeout can be broken down into two parts: substantial completion, and final completion and recommendation for acceptance. In addition, the recommendation for acceptance includes requirements for time analysis, contractor claims, claims against the contractor and withholdings, record documents, instruction and operation maintenance data, material parts and keys, changes, omissions and defects, certification of completion, final payment, warranties and guaranties, and testing and startup. The following procedures will outline the steps to be taken by the project staff.

Substantial Completion

§ 9.4 When the work is substantially complete, the contractor is usually required to submit to the construction manager the following items: a written notice stating that the work is substantially complete; a detailed and comprehensive list of all items to be completed or corrected; and the certification that the equipment has been tested, is operational, and that the required training has been provided.

After receipt of the contractor's notification, the construction manager or project manager will set up a substantial completion inspection to determine whether the project is ready for the punchlist inspection. Should the field engineer determine that the work is incomplete and does not warrant a punchlist inspection, the field engineer will notify the contractor in writing that the work is incomplete and require the contractor to promptly remedy the deficiencies and to provide a second notice of substantial completion.

When the work is ready for the punchlist inspection, the field engineer will arrange for inspection by owner personnel and representatives of the architect/engineer as necessary.

The construction management team and representatives of the architect/engineer will prepare a coordinated punchlist (see Figure 9-1) and will determine which items are to be completed by the contractor to achieve substantial completion. The list should be precise, giving all information required to locate and correct the deficient work.

Example

> WRONG: Unit heater in garage not installed as specified.
>
> RIGHT: Unit heater #2 at column line D-66, garage, dwg. 9-H-2, does not have vent and drain lines as required. Install as required.

The field engineer will then transmit the punchlist to the contractor, listing all omissions and defects and noting the items that must be complete to achieve substantial completion.

Once the punchlist items have been completed, the field engineer will prepare a certificate of substantial completion (see Figure 9-2) and will attach a list containing the balance of the punchlist items to be completed for the final completion. Other items which do not conform to contract documents may be added to the list at any time when they are found.

Final Completion and Recommendation of Acceptance

§ 9.5 The construction manager initiates the contract final completion acceptance recommendation (see Figure 9-3), which should contain the following items:

- schedule analysis and recommendation of any liquidated damages;

- list of all unresolved contractor claims;
- all claims and withholdings against the contractor;
- list of all record documents received from the contractor and copies of transmittals to the owner;
- list of all training and operations and maintenance manuals received and copies of transmittals to the owner;
- list of all materials, parts, and keys received and copies of transmittals to the owner;
- a copy of the change order report;
- a copy of the punchlist containing all completed omissions and defects and a note on any withholdings which are being retained for incomplete work;
- certification of substantial completion and final completion;
- recommendation for a final payment including withholdings, liquidated damages, and release of liens;
- a summary list of all warranties and guarantees and a copy of each; and
- a copy of all equipment test reports.

The acceptance recommendation report will be signed by the project manager. The construction manager will process the remaining field change orders. Once the field change orders are processed, the construction manager will deduct the appropriate amounts for withholdings, issue a final deductive change order, and release the remaining contingency with the final payment. The construction manager will process the certification of completion and issue to the contractor the final voucher noting the final deductions. Upon the receipt of the final voucher, the contractor usually has a specified time for initiating any final claim action concerning the withholdings and/or other unresolved claim issues.

FINAL PAYMENT

§ 9.6 After the completion certificate is accepted by the owner and the release of liens has been received, the field engineer will monitor all outstanding nonphysical work items, such as manuals, training, as-builts, and warranties, until they are completed. Upon completion of all contract items or the development of credit changes, the final payment will be initiated.

The final payment is included in the acceptance recommendation package and will contain deductions for withholding for incomplete work and liquidated damages. It will also account for all final change orders and release of retentions, and include the following:

- the total earned to date;
- the liquidated damages amount;
- the backcharge credit change order withholdings;
- the recommended retention release amount; and
- the recommended payment amount.

PUNCHLIST LOG

PAGE ______OF______

PROJECT ______________________________

CONTRACTOR ______________________________

PRELIMINARY INSPECTION DATE ____________________

LOCATION / STRUCTURE ____________________

INSPECTED BY ______________________________

SPEC./ DWG. REF.	DEFICIENCY ITEM	PERCENT COMPLETE/DATE				DATE RESOLVED
		1ST RE-INSP		2ND RE-INSP		
		%	DATE	%	DATE	

Figure 9-1

CERTIFICATE OF SUBSTANTIAL COMPLETION

TO:____OWNER

DATE OF SUBSTANTIAL COMPLETION:

____

PROJECT OR SPECIFIED PART SHALL INCLUDE:

____

____

____

____

PROJECT TITLE ________________________

PROJECT NO. ________________________

LOCATION ________________________

OWNER ________________________

CONTRACTOR ________________________

CONTRACT FOR ________________________

CONTRACT DATE ________________________

The work performed under this contract has been inspected by authorized representatives of the Owner, Contractor, and Architect/Engineer, and the Project (or specified part of the Project, as indicated above) is hereby declared to be substantially completed on the above date.

DEFINITION OF SUBSTANTIAL COMPLETION

The date of substantial completion of a project or specified area of a project is the date when the construction is sufficiently completed, in accordance with the contract documents, as modified by any change orders agreed to by the parties, so that the Owner can occupy or utilize the project or specified area of the project for the use for which it was intended.

A tentative list of items to be completed or corrected is appended hereto. This list may not be exhaustive, and the failure to include an item on it does not alter the responsibility of the Contractor to complete all the work in accordance with the contract documents.

________________________________ BY ________________________________

ARCHITECT/ENGINEER AUTHORIZED REPRESENTATIVE DATE

The Contractor accepts the above Certificate of Substantial Completion and agrees to complete and correct the items of the tentative list within the time indicated.

________________________________ BY ________________________________

CONTRACTOR AUTHORIZED REPRESENTATIVE DATE

The Owner accepts the project or specified area of the project as substantially complete and will assume full possession of the project or specified area of the project at _______(time), on _______(date). The responsibility for heat, utilities, security, and insurance under the contract documents shall be as set forth under "Remarks" below.

________________________________ BY ________________________________

AUTHORIZED REPRESENTATIVE DATE

REMARKS:

Figure 9-2

CONTRACT FINAL COMPLETION REPORT AND ACCEPTANCE RECOMMENDATION

TO			DATE
			CONTRACT NUMBER
PROJECT		CONTRACTOR	
ORIGINAL	CONTRACT AMOUNT	DURATION	LIQUIDATION DAMAGES
REVISED	CONTRACT AMOUNT		$__________PER DAY
STARTING DATES	CONTRACT	ACTUAL	NUMBER OF DAYS OVERRUN
COMPLETION DATES	CONTRACT	ACTUAL	TIME EXTENSIONS

1. TIME ANALYSIS SUMMARY AND LIQUIDATED DAMAGE ASSESSMENT
2. SUMMARY OF POTENTIAL CLAIMS FROM THE CONTRACTOR
3. SUMMARY OF POTENTIAL CLAIMS AGAINST THE CONTRACT AND/OR WITHHOLDINGS
4. RECORD DOCUMENTS
5. TRAINING, OPERATIONS AND MAINTENANCE MANUALS
6. MATERIALS, PARTS AND KEYS
7. CHANGES
8. OMISSIONS AND DEFECTS
9. CERTIFICATE OF COMPLETION
10. FINAL PAYMENT
11. WARRANTIES AND GUARANTEES
12. TESTING AND STARTUP
13. COMMENTS

THE CONTRACTOR IS COMPLETE AND IT IS RECOMMENDED THAT THE CONTRACT BE ACCEPTED

CONSTRUCTION MANAGER	CHIEF INSPECTOR

Figure 9-3

GLOSSARY OF CONSTRUCTION TERMS

by

John A. Ricchini, RA, CSI

INTRODUCTION

Words and phrases defined in this glossary are those frequently used in construction contracts and in some statutory or legal principles affecting the construction process. Definitions and explanations come from several sources and are composed to give the reader a practical understanding of how an intricate legal term or phrase applies to everyday construction concerns. Sources include *Black's Law Dictionary*, the *Manual of State and Local Construction Law* (American Bar Association National Institute, 1978), *Studies in Contract Law* (Murphy and Speidel, 2nd Edition 1979), *Construction Law Symposium* (St. Louis University Law Journal, 1979), *Legal Aspects of Architecture, Engineering and the Construction Process* (Justin Sweet, West Publishing 1970), and *Construction Law from a Practical Point of View* (John A. Ricchini, Professional Communications Institute 1979).

These definitions and explanations are not intended to give legal advice to the reader. If legal advice is needed, the reader should employ legal counsel. The author and the publisher will not be responsible for the reader's use of any of these definitions.

GLOSSARY

ACCELERATION COST—cost incurred by a contractor when the project is interfered with by the owner in such a way that the contractor must employ more manpower or work more hours in order to complete the project on time. If the contractor contributes to the cause of its own delays, acceleration cost may not be granted.

ACCEPTANCE—act of a person to whom a thing is offered by another whereby he receives the thing with the intention of retaining it, such intention being evidenced by a sufficient act. (See CONTRACT)

ACTIVE INTERFERENCE—action by a party to a contract which causes the other party of the contract to not complete the work of the project on time or in the manner established by the contract writing. Positive action must be performed on the part of the interfering party as opposed to passive negligence, which is inactive, permissive, or submissive.

ACTUAL DAMAGES—(ACTUAL LOSS)—damages resulting from real and substantial loss, as opposed to those which are merely theoretical, estimated, or anticipated. Actual damages represent the real and true value of the total loss suffered, as opposed to liquidated damages, which represent an estimated amount calculated as anticipated loss at a future time.

ADDENDA—modifications to the contract documents issued during the bid period. Addenda become official parts of the contract documents and are legally binding to the signatorees of the contract.

ADVERSARY—two parties to a contract are in an adversary or arms-length relationship to one another as a result of the commitment they have made to each other in the contract terms and conditions. This relationship is recognized by the courts and binds the two parties together in that relationship. In layman's language, it can be considered a relationship of mistrust.

AGENT—a person authorized by another to act for him or her; one who is employed to represent another in business and legal dealings with third persons. In a typical agency relationship, three parties are involved: a principal, an agent, and a third party. The agent represents the principal in dealing with the third party or parties. In the construction industry, a typical misunderstanding is that the architect is the agent to the owner in dealing with the third-party contractor. The architect, in a typical contract, is the representative of the owner and not of the agent. In some contracts, the construction manager is an agent of the owner. An agency relationship is established in writing (express agency) with all three parties acknowledging the relationship. An agency relationship may also be established by acts and/or omissions of the parties (implied or apparent agency) which will bind the parties legally in the same manner as an expressed agency relationship.

ALLOWANCE—a sum of money set aside by the owner to remove a particular portion of work from competitive bidding. This is typical of government-subsidized institutions with work that must be competitively bid and with projects in which certain portions of the work are proprietary and, therefore, must be removed from competitive bidding.

ALTERNATE—a material or method used in place of the base material or method specified for the project. In a typical construction contract, the owner chooses the alternate or remains with the base requirement, giving it control over the total cost of the project. An alternate differs from an option in that cost is a factor in the selection of an alternate by the owner, whereas an option does not have cost as a factor and the choice is made by the contractor. (See OPTION)

AMBIGUITY—doubtfulness; doubtfulness of meaning, duplicity, indistinctness, or uncertainty of meaning of an expression used in a written instrument. The courts, interpreting a writing, will permit parol evidence to clarify the writing if the writing is in fact ambiguous. However, the courts will not permit parol evidence if the writing is clear, even though it may be in error. (See PAROL EVIDENCE)

ANTICIPATORY BREACH—(ANTICIPATORY REPUDIATION)—established when a contractor makes a positive and unequivocal statement that it will not or cannot substantially perform the contract or when a contractor, by any voluntary affirmative act, renders substantial performance of its contract apparently impossible. Based on these two conditions, the owner may terminate the contract immediately or upon completion of a waiting period to determine the contractor's performance according to the contract writing. In either case, the owner must establish that the contractor's statement is positive and unequivocal. If the owner terminates the contractor for default after a statement which is ambiguous, the owner will be held to have wrongfully defaulted the contractor.

ANTICIPATORY REPUDIATION—(See ANTICIPATORY BREACH)

ANTITRUST LAWS—federal and state statutes to protect trade and commerce from unlawful restraints and monopolies. In the construction industry, bid rigging is considered a violation of antitrust laws. Those found guilty of bid rigging are assessed treble damages. (See BID RIGGING)

APPARENT AGENCY—an agency relationship created by an act of the parties and deduced from proof of other facts. (See AGENCY)

ARBITRATION—the submission of a dispute to a third party (individual or panel), known as arbitrator(s), whose judgment is final and binding. Decisions at arbitration hearings, unlike those in judicial cases, do not establish precedents.

ARBITRATOR—one who resolves disputes between two parties. In a typical construction contract, the architect is designated as an arbitrator in resolving the disputes between the owner and the contractor. Unlike formal arbitration (as established by the American Arbitration Association), an architect acting as arbitrator in the construction process is the first level for resolving disputes, and its decision is not final and binding.

ARCHITECT—the person or organization hired by the owner to design the project. The architect's duties consist primarily of the production of the plans and specifications from which the building will be constructed. The architect may also preside at the bid opening, monitor the construction process to assure that the owner's interests are protected, and approve payments to the contractor. Its relationship to the owner is that of an independent contractor. All architects must be licensed by the states in which they practice. In addition to the contract with the owner, the architect also will enter into contracts with consultants (structural, mechanical, electrical engineers,etc.) but will not execute a contract with the contractor.

ASSIGNMENT—a legal action which allows a person who is not party to a contract to obtain the contract rights of a party who is. A contractor, for example, may assign the rights contained in its contract with the owner to a subcontractor. In a similar manner, the architect can assign portions of the design of the project to its consulting engineers, primarily in the areas of structural, mechanical, and electrical design.

ATTACHMENT—the act or process of taking, apprehending, or seizing person or property by virtue of a writ, summons, or other judicial order and bringing the same into the custody of the law; a remedy ancillary to an action by which the plaintiff is enabled to acquire a lien upon the property or effects of the defendant for satisfaction of judgment which the plaintiff may have obtained. (See LIEN)

BETTERMENT—an improvement brought upon an estate (land and/or buildings) which enhances its value more that mere repairs. The improvement may either be temporary or permanent. This term also applies to denote the additional value which an estate acquires in consequence of some public improvement, such as the widening of a street, etc.

BID—an offer to perform a contract for work and labor or for supplying materials at a specified price. In the construction industry, a bid is considered an offer by the contractor to the owner. A bid, as an offer, becomes a contract once the owner accepts the bidder's offer with all other contractual requirements in order. (See CONTRACT)

BID, UNIFIED —(See UNIFIED BID)

BID BOND—(See BOND)

BID DEPOSITORY—a clearing house for subcontractors to submit their bids for a particular project and for prime contractors to receive bids from the various subcontractors. In California, a bid depository was found in violation of antitrust laws based on its rules for membership imposing fine, suspension, or expulsion to members not abiding by the rules.

BID INVITATION—(See INVITATION TO BID)

BID REJECTION—the act of not allowing a bid to stand because of an impropriety in the process of submission or as a result of the owner's arbitrary decision to reject the bid. The owner, in a typical contract, reserves the right to reject any and all bids. However, in rejecting a bid, an owner and its architect run the risk of interfering with the bidder's right to do work or of defamation of character on the part of the bidder.

BID RIGGING—(See ANTITRUST LAWS)

BID SECURITY—(See BOND)

BID SOLICITATION—(See INVITATION TO BID)

BOARD OF CONTRACT APPEALS—an independent administration quasi-judicial board to decide all public contract disputes. Various states have created these boards to relieve the courts from the backlog of cases related to public contracts. Note that these boards hear only disputes related to public contracts and not to private contracts.

BOILER PLATE—a term used to represent standard legal conditions inserted at the "front end" of a construction contract. These conditions are typically titled "General Conditions," "Supplemental Conditions," and/or "Special Conditions" and are inserted at the front end of the project manual.

BOND—an instrument with a clause, with a sum fixed as a penalty, binding the parties to pay the same, and with the condition that the payment of the penalty may be avoided by the performance of certain acts by some, one, or more of the parties; a certificate or evidence of a debt; a mere promise to perform or pay; a written obligation. In the construction industry, there are several types of bonds, including bid bonds, performance bonds, and payment bonds. A bid bond is a form of security to insure that the bidder will enter into the contract if the award is made to it. A performance bond insures completion of the project by the contractor, guaranteeing that if the contractor defaults, the bonding company will step in and finish the work. A performance bond also is applicable between a prime contractor and its subcontractor, assuring the prime that the subcontractor will perform or pay. A payment bond (sometimes known as a labor and material payment bond) provides a source of payment for the contractors' or subcontractors' labor and materialmen.

BUILDER—one whose occupation is the building or erection of structures, the controlling and directing of construction, or the remodeling and adapting to particular uses of buildings and other structures. The term "builder" is sometimes used interchangeably with the word "contractor." (See CONTRACTOR)

BUILDING CODE—there are several model codes, including Southern Standard Building Code (SSDSC), Uniform Building Code (UBC), Building Officials and Code Administrators (BOCA), and the National Building Code (NBC), one of which is enacted in most jurisdictions. A code is not applicable in a certain jurisdiction or locality until it is enacted (legislated) into local law.

CAPACITY—the attribute of persons which enables them to perform civil or juristic acts; necessary for parties entering into a contract. (See CONTRACT)

CASE LAW—the aggregate of reported cases forming a body of jurisprudence or the law of a particular subject as evidenced or formed by the adjudged cases; distinct from statutes and other sources of written law.

CAVEAT—a caution; literally, "let him beware."

CERTIFICATE—a written assurance, or official representation, that some act has or has not been done, that some event occurred, or that some legal formality is being complied with; a written and signed document establishing that a fact is true.

CERTIFICATE OF OCCUPANCY—a document issued by the building inspector certifying that the structure conforms to all relevant code sections and is, therefore, safe for use. An owner must obtain a certificate of occupancy before he or she can use a building. A new building cannot be considered complete until a certificate of occupancy has been issued. In some instances, a partial certificate of occupancy will be issued for portions of the building to be occupied.

CERTIFICATE OF PAYMENT—a document issued by the architect in which the architect certifies that the contractor has adequately performed. The certificate is then presented to the owner for payment to the contractor.

CERTIFICATE OF SUBSTANTIAL COMPLETION—the document issued by the architect when the building, or a portion thereof, is complete to the degree that the owner can use the building, or a portion thereof, for its intended purpose. (See SUBSTANTIAL COMPLETION)

CHANGE—a revision to the original contract documents. A change differs from a modification in that the modification is agreed to by both parties of the contract; however, a change may be made unilaterally by the owner in spite of the contractor's lack of agreement.

CHANGE ORDER—a document issued by the architect directing the contractor to erect some portion of the building in a manner different than described in the original plans and specifications. This change must have an effect on the price and/or time of the contract in order to constitute a change order. If the price and/or time is not affected, then the change is a field order or minor change order and not a change order. The change may be requested by the architect, owner, or contractor.

CLAIM—a demand, an assertion, a pretense, a right or title to. An action initiated by one of the parties of a contract against the other party. This action may be in the form of a written letter, a legal document, or some instrument establishing the difference between the two parties. (**NOTE**: A letter is sufficient, in the eyes of some courts, to establish a claim.)

CLOAK OF IMMUNITY—legal status granted to an architect in the quasi-judicial role as arbitrator in settling a dispute between the owner and the contractor. This cloak protects the architect from liability by either party (owner or contractor) as a result of the decision rendered in resolving the dispute. (See IMMUNITY)

CODE—(See BUILDING CODE)

COLLUSION—an agreement between two or more persons to defraud a person of his or her right by the forms of law or to obtain an object forbidden by law; a secret combination, conspiracy, or concert of action between two or more persons for fraudulent or deceitful purposes.

COMPENSATORY DAMAGES—(See DAMAGES)

COMPETITIVE BIDDING—a process whereby sealed proposals are submitted to the owner for consideration. Competitive bidding is mandatory on public works projects. A private owner may choose to use competitive bidding in securing the most economical contractor for the construction of the project. However, a private owner is not legally bound to the competitive bidding process.

CONSEQUENTIAL DAMAGES—(See DAMAGES)

CONSIDERATION—the inducement to a contract; the cause, motive, price, or impelling influence which induces a contracting party to enter into a contract; the reason or material cause of a contract. (See CONTRACT)

CONSTITUTION—the written instrument agreed upon by the people of the United States, or of a particular state, as the absolute rule of action and decisions for all departments and officers of the government in respect to all the points covered by it. This instrument must control until it is changed by the authorities which established it. Any act or ordinance of any government department or office opposed to it is null and void. Several states have enacted statutes which have affected the construction industry and have been found unconstitutional or null and void in their application. One such statute is the statute of limitations which is applied for the protection of the owner and architect but not for the contractor.

CONSTRUCTION MANAGEMENT—a process of professional management applied to a construction program from conception to completion for the purpose of controlling time, cost, and quality. Ideally, the construction management organization links itself to the owner as an agent and thereby places itself in a fiduciary relationship with the owner. In this relationship, the construction manager can properly represent the owner to both the design professional and the contractors without concern regarding conflict of interest on his part.

CONSTRUCTIVE—that which has the character assigned to it in its own essential nature but acquires such character as a consequence of the way in which it is regarded by a rule or policy of law; hence, inferred, implied, or made out by legal interpretation. The term "constructive" typically is used with other legal terms such as "acceleration," indicating that in the absence of an acceleration clause, it is the actions of party that determine the validity of acceleration costs. Another application is in the use of the term "constructive change," indicating that although a change may not have been directed, it is implied by the act or omission of the parties involved. (See ACCELERATION and CHANGE)

CONTINUOUS TREATMENT—an uninterrupted, unbroken series of activities or events. This theory is sometimes employed in the determination of statute of limitation claims regarding the commencement of the time for the claim. The statute of limitation typically starts to run upon completion of the project. However, if the contractor is required to repair defects in the work and, as a result, renders "continuous treatment" to the work, the contractor may extend the time for commencement of the statute.

CONTRACT—a promissory agreement between two or more persons that creates, modifies, or destroys a legal relationship. Several essential elements must be present in order to render a contract valid. These elements include an offer, acceptance, and consideration on the part of both parties, the capacity of both parties to contract, a state of mind (mutuality of assent), and the "meeting of minds." In the construction industry, especially in public bidding, the bid proposal is considered an offer and the owner's selection of the bid is the acceptance. Consideration is the giving up of something on the part of both parties (the owner gives money while the contractor gives labor, material, etc., in the construction process). The capacity of both parties represents their legal standing in relation to one another, namely as legally recognized principals of the organizations entering into the contract. The state of mind (mutuality of assent) of the individuals must be such that they are free to enter into the contract or not to enter into the contract. If coercion is present, then the contract could be rendered null and void. The meeting of the minds represents that which was intended by both parties at the time of the signing of the contract and that both parties were in harmony with each other's intentions. (See BID)

CONTRACTOR—anyone who contracts to provide the labor and services necessary to complete a project. A contractor may be hired by the owner or by another contractor. When the contractor is hired directly by the owner, the contractor is classified as a prime contractor. When a contractor is hired by another contractor, the contractor is classified as a subcontractor in relation to the project. (See SUBCONTRACTOR)

CONTRACTUAL DUTY—(CONTRACTUAL OBLIGATION) the obligation which arises from a contract or agreement. In a typical contract agreement, the parties are required to fulfill the duties enumerated in the contract writing between the two parties, but also from the contract agreed to by other parties. An example of this is the duty owed by the architect to the contractor as a result of the requirements called out in the contract between the owner and the contractor.

CONTRIBUTION—the sharing of a loss or payment among several debtors. The act of any one or several of a number of co-debtors in reimbursing one of their number which has paid the whole debt or suffered the whole liability, each to the extent of its proportionate share; The right of one who has discharged a common liability to recover from another, who is also liable, the portion which he or she ought to pay or bear. In many jurisdictions, the damages will be assessed to the parties held liable based on their contribution to the negligence.

CUSTOM—a usage or practice of the people which, by common adoption and acquiescence and by long and unvarying habit, has become compulsory and has acquired the force of a law with respect to the place or subject matter to which it relates. On the technical side of the construction industry, this term can apply to techniques and methods of construction, such as the finishing of a concrete slab with a trowel. Administratively, it is the custom of an architect to monitor the construction phase of the work, unless the writing contains a clause deleting that requirement.

DAMAGES—compensation for a loss or injury suffered; compensation which may be recovered in the courts by any person who has suffered loss, detriment, or injury, whether to his or her person, property, or rights, through the unlawful act, omission, or negligence of another. In the courts there are many divisions pertaining to damages which cannot be covered here.

DAMAGES, ACTUAL—real, substantial, and just damages, or the amount awarded to a complainant in compensation for actual and real loss or injury, as opposed to "nominal" or "punitive" damages.

DAMAGES, COMPENSATORY—repair or replacement of the loss caused by the wrong or injury and nothing more.

DAMAGES, CONSEQUENTIAL—such damage, loss, or injury which does not flow directly from the act of the party but only from some of the consequences or results of such act.

DAMAGES, DELAY—the economic loss suffered as a result of extended time from that of the original time stipulated in the contract writing. This differs from property damage and personal damage.

DAMAGES, LIQUIDATED—a specific sum of money expressly stipulated by the parties to a bond or contract as the amount of damages to be recovered by either party for a breach of the agreement by the other. In the construction industry, it is an amount established in the contract writing to be withheld by the owner on a daily basis for every day past the stipulated completion date of the contract. A "liquidated damages" clause is to fix the amount to be paid in lieu of performance. "Penalty" clauses, without some kind of balancing bonus, are rendered unenforcable in the courts of law.

DAMAGES, PUNITIVE—awarded by the courts in the amount of three times the actual damage. Treble damages usually apply in antitrust actions.

DEFAULT—an omission of that which ought to be done; a failure to perform a legal duty.

DELAYED DAMAGES—(See DAMAGES)

DEMURRER—an allegation of a defendant which, admitting the matters of fact alleged by the bill to be true, shows that they are insufficient for the plaintiff to proceed upon or to oblige the defendant to answer.

DEPOSITORY, BID—(See BID DEPOSITORY)

DESIGN-BUILD—a method of organizing a building project in which a single entity undertakes the design and erection of the structure at a set fee negotiated in advance. Unlike the conventional construction contract whereby an owner hires both an architect and a contractor separately, in the design-build contract, the owner negotiates only one contract with one organization.

DEVIATION—a change made in the progress of a work from the original terms, design, or method agreed upon.

"DIFFERING SITE CONDITIONS" CLAUSE—("CHANGED CONDITIONS" CLAUSE)—typically provides that in the event that the physical conditions at the site of the work vary materially from those represented or reasonable anticipated and in a manner which increases the time or cost of performance, the contractor is entitled to additional compensation or an extension of time.

DISCLAIMER—the disavowal, denial, or renunciation of an interest, right, or property imputed to a person or alleged to be his; also the declaration, or the instrument, by which such disclaimer is published.

DISCLOSURE—to bring into view by uncovering, to lay bare, to reveal, to free from secrecy or ignorance, or to make known; revelation; the impartation of that which is secret; that which is disclosed or revealed.

DISCOVERY—the ascertainment of that which was previously unknown, the disclosure or coming to light of what was previously hidden, the acquisition of notice or knowledge of given acts or facts as in regard to the discovery of fraud affecting the running of the statute of limitations, or the granting of a new trial for newly discovered evidence; disclosure of facts resting in the knowledge of the defendant or of deeds, writings, or other things in his custody or power.

DOCUMENT, DOCUMENTATION—instruments which record, by means of letters, figures, or marks, matter which may be evidentially used; the deeds, agreement, title papers, letters, receipts, and other written instruments used to prove the facts.

DUTY, CONTRACTUAL—(See CONTRACTUAL DUTY)

ECONOMIC LOSS—additional cost incurred by an individual other than property damage or personal injury. In the construction industry, an economic loss may be represented by a loss in profits or a loss due to a delay in the contractor's schedule. (See DELAY DAMAGES)

ENGINEER—a person with a particular expertise in a limited area of building design. An engineer typically may specialize in structural, mechanical, electrical, or plumbing design. It is the limitation of this speciality which distinguishes an engineer from the architect, who has general responsibility for the entire project. Engineers ordinarily are hired as consultants to assist the architect. The contract between the architect and the engineer usually reflects the same terms and conditions that exist in the contract between the owner and the architect. In some instances, such as an industrial project, the roles are reversed in that the owner hires the engineer as the prime designer, and the engineer, in turn, hires the architect as a consultant for the building enclosure.

EQUITABLE DOCTRINE—just and conformable to the principles of justice and right; existing in equity; available or sustainable only in equity or only upon the rules and principles of equity. (See EQUITY, COURT OF)

EQUITY, COURT OF—court which administers justice according to the system of equity and according to a peculiar course of procedure or practice. Equity denotes the spirit and habit of fairness, justness, and right dealing which should regulate the interaction of men. Its obligation is ethical rather than jural. It is grounded in the precepts of the conscience, not in any sanction of positive law. It is justice that is ascertained by natural reason or ethical insight independent of the formulated body of law.

ESTOPPEL, PROMISSORY—an equitable doctrine which holds the promisor bound to a promise if injustice can be avoided only by enforcement of the promise. A typical application of this doctrine in the construction industry is holding a subcontractor to its bid submitted to the prime contractor.

EXCLUSIVITY OF CONTRACT PROVISIONS—when a remedy for breech is included as a part of the contract, that remedy is considered exclusive of other remedies provided by law. Some courts do not recognize the exclusivity of a contract provision unless it is specifically stipulated that the remedy is exclusive. Courts typically will look at all of the facts and circumstances surrounding the agreement as a means of determining the intention of the parties and will refuse to exclude other remedies unless such a result is required by a consideration of the facts of the particular agreement.

EXCULPATORY LANGUAGE—clause in which a party who may suffer a loss agrees not to institute legal action against the party who may cause the loss. The classic example is a patient who, upon entering the hospital, agrees not to institute any legal action against the hospital or any of the doctors in the event he suffers injury or death because of an act of the hospital or the doctors. In the jargon of the construction industry, indemnification clauses and disclaimer clauses are considered exculpatory language.

EXPERT WITNESS—may be a person of science, one educated in the arts, or a person possessing special or peculiar knowledge acquired from practical experience.

EXPRESS AGENCY—(See AGENCY)

EXPRESSED WARRANTY—in contract and sales, a promise created by the apt and explicit statements of the seller or person to be bound. (See WARRANTY)

FAST-TRACK METHOD—a way of organizing a design program which allows the contractor to begin construction on earlier phases of the project before the plans are completed for the entire project. Caution must be exercised in the signing of a contract using these fast-track methods because of the lack of information typically expressed in a conventional method of contract. Many changes may result when going from phase to phase, and provisions must be included in the contract to compensate the contractor for additional work.

FIDUCIARY—a person holding the position of a trustee or character analogous to that of a trustee in respect to the trust and confidence involved and the scrupulous good faith and candor which are required. In the construction industry, the architect and the owner are in a fiduciary relationship in respect to the contractor.

FIELD ENGINEER—an engineer assigned to a project during the construction phase and located at the project on a full-time basis.

FIELD ORDER—a document issued by the architect directing the contractor to erect some portion of the building in a manner different from that described in the plans and specifications. A field order is issued when the modification will not affect the money and/or the time spent on the project. These factors distinguish a field order from a change order. The change may be requested by the architect, owner, or contractor. (Field orders sometimes are known as minor change orders.) (See CHANGE ORDER)

FINALITY OF DECISION—a contract provision or the procedure of a legally recognized process which states that the decision rendered in the settlement of a dispute is final. Pursuant to such a provision, the courts will accord finality to that decision absent gross error for arbitrary and capricious action. In the arbitration process, the decision rendered by the arbitrators is final.

"FLOW-DOWN" CLAUSE—in a contract between a subcontractor and a prime contractor, the performance of the subcontractor will be tied to the prime contractor in the same manner as the prime's performance is tied to the owner. Some contracts between contractors and subcontractors require more of the subcontractor than is required of the prime contractor by the owner. In a "flow-down" clause, the same requirements are established as a minimum requirement for the subcontractor.

FOREIGN CORPORATION—an organization not incorporated in the state or jurisdiction in which it is performing work. A contractor must meet the legal requirements of the state in which it is performing work. These requirements may include incorporation and licensing as a construction contractor in that state.

FOUR-CORNER RULE—the face of a written instrument. That which is contained on the face of a deed, without any aid from the knowledge of the circumstances under which it is made, is said to be within its four corners. In the construction industry, the contract documents, including the drawings, specifications, general conditions, etc., form the face value of the contract.

FRAUD, FRAUDULENT—an intentional perversion of truth for the purpose of inducing another, in reliance upon it, to part with some valuable thing or to surrender a legal right; a false representation of a matter of fact, whether by words, by conduct, by false or misleading allegations, or by concealment of that which should have been disclosed, which deceives and is intended to deceive another so that he or she shall act upon it to his or her legal injury.

GENERAL CONDITIONS—those portions of the contract documents which define, set forth, or relate to contract terminology, the rights and responsibilities of the contracting parties and of others involved in the work, and similar provisions of a general nontechnical nature. Conditions can be either expressed, which are stated in the contract, or implied, which are not set forth in words but arise out of the intentions of the parties to the contract.

GENERAL CONTRACTOR—the builder of the portion of the building which is considered the general portion or the architectural portion. This terms sometimes is erroneously interchanged with the term "prime contractor." (See PRIME CONTRACTOR)

GUARANTY—a collateral agreement for performance of another undertaking; a promise to answer for payment of debt or performance of an obligation if the liable person fails to make payment or perform the obligation.

"HOLD-HARMLESS" CLAUSE—this heading is interchangeably used with the heading "Indemnification Clause." (See INDEMNIFICATION)

IMMUNITY—exemption from performing duties which the law generally requires others to perform. (See CLOAK OF IMMUNITY).

IMPLIED—where the intention is not manifested by an explicit and direct word but is gathered by implication or deduction from the circumstances.

IMPLIED AGENCY—an agency relationship created by acts of the parties and deduced from proof of other facts. (See AGENCY)

IMPLIED CONTRACTUAL PROVISIONS—provisions which do not appear in the written embodiment of the agreement, but which exist by implication. These primarily include the implied duties to cooperate and to disclose, the implied warranty of specification suitability, and the implied covenant of good faith and fair dealing. Recovery under these implied clauses may not be subject to the limitations on recovery under the expressed provisions of the contract.

IMPLIED WARRANTY—a promise established by implication or inference from the nature of the transaction or the situation or circumstances of the parties. (See WARRANTY)

IMPOSSIBILITY OF PERFORMANCE—a requirement of the contract which is physically impossible to perform within the existing state of the art. Three factors must exist to render a requirement impossible: (1) the impossibility must be inherent in the nature of the act to be performed rather than personal to the contract, (2) the facts which make the performance impossible must not have been forseeable, and (3) the person seeking to be excused from performance must have been in no way responsible for the impossibility.

IMPRACTICABILITY, COMMERCIAL—the doctrine that recognizes that, in some instances, contract performance may become so costly that its impracticability makes it the equivalent of impossibility. (See IMPOSSIBILITY OF PERFORMANCE)

IMPUTED KNOWLEDGE—knowledge of a fact which is attributed vicariously to another. Knowledge is said to be imputed to a person when it is ascribed or charged to the person not because he or she is personally cognizant of the fact or responsible for it, but because another person, over whom the first person has control or for whose acts or knowledge he or she is responsible, is cognizant of it or responsible for it. In an agency relationship, the principal has knowledge imputed to him or her when the agent receives or is made cognizant of that knowledge. (See AGENCY)

INCORPORATED PAPERS—where the signatorees execute a contract which refers to another instrument in such a manner as to establish that they intended to make the terms and conditions of that other instrument a part of their understanding. The two instruments may be interpreted together as the agreement of the parties.

INDEMNIFICATION—the process by which one party seeks to protect itself from any claims by a plaintiff who has been injured or who has suffered loss. One method of obtaining indemnification is to obtain a promise from the contractor that it will insure the owner, and in some cases the architect, against any liens or suits by a third party not privy to the contract. The courts generally enforce contractual indemnification provisions; but, they are hesitant to permit a party to use indemnification when that party has played a major role in causing the loss. Indemnification is a contractual obligation by which one person or organization agrees to secure another against loss or damage from specified liability.

INJUNCTION—a prohibitive writ issued by a court of equity to a party defendant, forbidding the latter to do some act or to permit its servants or agent to do some act which it is threatening or attempting to commit, or restraining it in the continuance thereof, such as being unjust, inequitable, or injurious to the plaintiff. In the application of the equitable doctrine of promissory estoppel, one can only stop the subcontractor from withdrawing its bid. This is an injunctive procedure preventing the subcontractor from performing an act, but it cannot assess damages against the subcontractor.

INSPECTION TEAM—the inspectors assigned to a project for the purpose of carrying out the quality control plan.

INSURANCE—a contract whereby, for a stipulated consideration, one party undertakes to compensate the other for loss on a specified subject by specifying perils. The party agreeing to make the compensation usually is called the insurer or underwriter; the other is the insured or assured; the agreed consideration is the premium; the written contract is the policy; the events insured against are risks or perils; and the subject, right, or interest to be protected is the insurable interest. Insurance is contract whereby one undertakes to indemnify another against loss, damage, or liability arising from an unknown or contingent event and is applicable only to some contingency or act to occur in the future.

INTERFERENCE—the act of hampering, hindering, disturbing, intervening, interposing, or taking part in the concerns and affairs of others. In the construction industry, when a contractor has work interrupted by the acts of the architect or owner, it may file suit on the grounds of interference. However, before liability will be assessed, most courts require that interference with the contract be intentional and not merely negligent.

INVITATION TO BID—a solicitation for competitive bids; an invitation to submit offers on behalf of contractors, which are then subject to acceptance by the procuring agency or owner to form the basis of the contract. The invitation to bid competitively is not an offer on behalf of the procuring agency or owner to contract but is simply a request or solicitation for offers to contract.

JUDICIAL—belonging to the office of a judge, as in a judicial authority, a court of justice, a judicial writ, or a judicial determination.

JUDICIARY—pertaining or relating to the courts of justice, to the judicial department of government, or to the administration of justice; that branch of government invested with the judicial power; the system of courts in a country.

LATENT—hidden, concealed, dormant; does not appear upon the face of a thing, as in a latent ambiguity.

LATENT/PATENT TEST—(See PATENT/LATENT TEST)

LIABILITY—bound or obliged in law or equity; responsible or answerable to make satisfaction, compensation, or restitution.

LICENSE—certificate or document which gives permission; a permission by a competent authority to do some act which, without such authorization, would be illegal or would be a trespass or a tort.

LIEN—a charge, security, or encumbrance upon property; a claim or charge on property for payment of some debt, obligation, or duty.

LIEN, MECHANIC'S—a claim created by law for the purpose of securing priority of payment of the price or value of the work performed and materials furnished in erecting or repairing a building or other structure and, as such, attached to the land as well as to buildings and improvements erected thereon. (See ATTACHMENT)

LIEN, PARTIAL WAIVER OF—in the construction industry, a document used to certify that a portion of the total amount due to a subcontractor has been paid and, therefore, that that portion or amount of money cannot be used as a basis for a lien against the property.

LIEN, WAIVER—to deny the right expressed in the lien. In the construction industry, it is a certificate issued upon completion of the work, signifying that all monies have been paid and that the right to lien against the property is removed.

LIMITATIONS, STATUTE OF—a statute prescribing limitations to the right to bring on action based on certain prescribed causes of action; that is, declaring that no suit shall be maintained on such causes of action unless brought within a specified period after the right has accrued; a certain time allowed by a statute for litigation. The provisions of state constitution are not a grant but are a limitation of legislative power.

LIQUIDATED DAMAGES—(See DAMAGES)

MANAGEMENT, CONSTRUCTION—(See CONSTRUCTION MANAGEMENT)

MANAGEMENT, PROJECT—(See PROJECT MANAGEMENT)

MANDAMUS—a writ issued from a court of superior jurisdiction and directed to a private or municipal corporation, or any of its offices, or to an executive, administrator, or judicial officer, commanding the performance of a particular act therein specified and belonging to its public, official, or ministerial duty or directing the restoration of the complainant to rights or privileges of which he or she has been illegally deprived; a command from a higher court to a lower court to perform a particular act. In the construction industry, a writ is issued to the contracting officer conducting a bid opening session or the letting of contracts if the officer is not complying with the proper legal procedures. If a public body is withholding the execution of a contract, mandamus may be applied to compel that body to act. (See MANDATE)

MANDATE—a precept or order issued by superior court upon the decision of an appeal or writ of error which directs action to be taken or disposition to be made of case. In some state jurisdictions, the term "mandate" has been substituted for "mandamus" as the formal title of that writ. (See MANDAMUS)

MANDATORY CLAUSES—(MANDATORY PROVISIONS)—clauses which must appear in the contract writing due to their legal status as a federal, state, or local law. The amount of minority business participation or the licensing of a contractor or subcontractor are clauses which fall into this category in certain jurisdictions.

MATERIAL VARIANCE—a deviation from that which was specified in the original contract documents. In the bid process, a material variance from that which is required in the bid documents will be the basis for rejection of the bid. The degree of variance in a bid process is determined by whether the bidder's proposal gives it an advantage or benefit not enjoyed by the other bidders. A mere irregularity in form which can be corrected upon the opening of the bid is not considered a material variance.

MECHANIC'S LIEN—(See LIEN)

MEETING OF MINDS—the "meeting of minds" required to make a contract is not based on secret purposes or intentions on the part of one of the parties, which it has stored away and not brought to the attention of the other parties, but must be based on purpose and intention which has been made known or from which all of the circumstances should be known. (See CONTRACT)

MERCHANTABILITY—the article sold will be of the general kind described and reasonably fit for the general purpose for which it shall have been sold. Where the article sold is ordinarily used in only one way, its fitness for use in that particular way is impliedly warranted unless there is evidence to the contrary.

MISREPRESENTATION—any manifestation by words or other conduct of one person to another that, under the circumstances, amounts to an assertion not in accordance with the facts. A party may be guilty of misrepresentation if it has erred in giving professional opinions or in making representations as to existing facts or conditions which a third party has relied upon in the performance of its work.

MUTUALITY OF ASSENT—compliance, approval of something done, or a declaration of willingness to do something in compliance with a request; an acting by two parties to perform a duty toward each other. (See CONTRACT)

NEGLIGENCE—failure to exercise the degree of care which a reasonable and prudent party would exercise under the same circumstances. Negligence is committed when a contractual duty is breached. A good example of negligence is where an architect failed to indicate in the plans the existence of an electric power line which he or she knew to be in the area of construction.

NO DAMAGE FOR DELAY—a clause contained in contracts which grants a party to the contract an extension of time but does not reimburse that party for any additional costs suffered during that time.

NULL AND VOID—naught, of no validity or effect. When used in a contract or statute, it often is construed as meaning voidable. A contract is rendered null and void when one of the essential elements that make up a contract is missing. An example of this is that when an organization is not licensed to perform work in a particular state, that organization does not have the capacity to execute contracts in that state. Such a contract can then be rendered null and void because of its deficiency regarding the capacity of one of the parties.

NULLITY—nothing; an act or proceeding in a cause which the opposite party may treat as though it had not taken place or which has absolutely no legal force or effect.

OFFER—an act on the part of one party whereby it gives to another the legal power of creating the obligation called contract; a proposal to do a thing; an element of a contract. It must be made by the party which is to make the promise, and it must be made to the party to which the promise is made. It may be made either by word or by signs, either orally or in writing, and either personally or by a messenger; but, in whatever way it is made, it is not an offer in law until it comes to the knowledge of the party to which it is made. An offer must be so definite in its terms, or require such definite terms in acceptance, that the promises and performances to be rendered by each party are reasonably certain. (See CONTRACT)

OPTION—a choice; the power or liberty of choosing; something that is or can be chosen. In the construction industry, an option is presented to the building contractor in the form of materials and/or methods which vary from from the base requirements, which it may choose in order to meet other requirements of the contract. An example would be to choose a method which would employ more minorities to meet the minority quota. An option has no effect on the cost to the owner. (See ALTERNATE)

O.S.H.A. (OCCUPATIONAL SAFETY AND HEALTH ACT)—a federal act creating an agency responsible for safety and health in the work place. The agency has the authority to issue citations to violators of the federal regulations imposed by the agency. There have been instances in the construction industry where O.S.H.A. has been used by the courts to establish a standard of care for the participants in the construction process.

OWNER—the party at the instance of which the project is undertaken and the one which will take title to it when it is completed; the party in which is vested the ownership, dominion, or title to property. On a construction project, the owner typically contracts independently with the architect or engineer and with the contractor.

PAROL EVIDENCE—oral or verbal evidence; that which is given by word of mouth; the ordinary kind of evidence given by witnesses in court. In a particular sense, and with reference to contracts, deeds, wills, and other writings, parol evidence is the same as extraneous evidence or evidence taken from outside of the contract writing.

PAROL EVIDENCE RULE—under this rule, when parties put their agreement in writing, all previous oral agreements merge in the writing and a contract, as written, cannot be modified or changed by parol evidence in the absence of a plea of mistake or fraud in the preparation of the writing. But, this rule does not forbid a resort to parol evidence not inconsistent with the matters stated in the writing. In common layman's terms, parol evidence or extrinsic evidence is not permitted to be used a part of the contract writing once the contract is executed. However, should the writing be ambiguous and in need of clarification, then the courts will permit parol evidence to be received concerning the contract writing. In the construction industry, only the contract is executed, the bid proposal cannot be entered as evidence contrary to the contract writing unless the contract writing is ambiguous and the bid proposal is needed for clarification of the ambiguity.

PARTIAL LIEN WAIVER—(See LIEN, PARTIAL WAIVER OF)

PATENT/LATENT TEST—determines whether the danger which caused the damage was latent (hidden) and, therefore, beyond the control of the observer or patent (readily seen upon a reasonable inspection) and, therefore, within the control of the observer. Application of this test to the construction industry is enforced when the building is turned over to the owner. If the danger can be observed at the time of the acceptance of the building by the owner, but the owner does not make the contractor aware of the deficiency, then the owner will be held responsible for any future damage. However, if the danger is latent and not observable by the owner, then the contractor will be held responsible for any future damage emanating out of this danger.

PAYMENT BOND—a legal instrument which provides a source of payment for labor and materialmen should their employer fail to pay them because of either default or bankruptcy. (See BOND)

PERFORMANCE BOND—a legal instrument which assures that if the contractor defaults, the surety company will complete performance or pay damages to the extent of the bond. (See BOND)

PLAINTIFF—a person or organization which brings an action; the party which complains or sues in a personal action and is so named on the record.

PRECEDENT—an adjudged case or decision of a court of justice considered as furnishing an example or authority for an identical or similar case arising afterward or for a similar question of law. It means that a principle of law actually presented to a court of authority for consideration and determination has, after due consideration, been declared to serve as a rule for future guidance in the same or analogous cases, but matters which merely lurk in the record and are not directly advanced or expressly decided are not precedent.

PRIME CONTRACTOR—the party signing a contract with another party to directly perform the work required by that contract. (See CONTRACTOR and SUBCONTRACTOR)

PRIVITY—relationship of a party which has any part or interest in any action, matter, or thing. Privity of contract is that relationship that exists between two or more contracting parties. In a typical construction project, the contractual relationship between the participants is one of privity between the owner and the design professional and the owner and the contractor. However, there is no privity or contract between the design professional and the contractor.

PRIVITY—(NO PRIVITY RULE)—in the 1800's, many cases were settled when the plaintiff was denied access to the bench due to the no privity rule (no contract existed between the plaintiff and the defendant). However, in recent decades the no privity rule has given way to the notion of third-party liability. (See THIRD-PARTY LIABILITY)

PROJECT MANAGEMENT—a system of organizing a construction project from conception to the completion of the project. This system includes management of the preparation of the contract documents, the bid process, and the construction phase. This term sometimes is interchangeably used with the term "construction management." (See CONSTRUCTION MANAGEMENT)

PROMISSORY ESTOPPEL—(See ESTOPPEL, PROMISSORY)

PROTEST—a formal declaration made by a party interested or concerned in some act about to be done, or already performed, whereby it expresses its dissent or disapproval or affirms the act against its will. The object of such a declaration generally is to save some right which would be lost to the party if its implied assent could be made out or to exonerate itself from some responsibility which would attach to it otherwise. In common jargon, a protest is considered the initial act in establishing a claim to retain a party's contractual rights.

PUNITIVE DAMAGES—relating to punishment; having the character of punishment or penalty; inflicting punishment or a penalty. (See DAMAGES)

QUALITY ASSURANCE—policy in regard to assuring that quality will be achieved on a program or project.

QUALITY ASSURANCE PLAN—a plan to implement the policies stated in the quality assurance statement of an organization.

QUALITY CONTROL—the implementation of the quality assurance plan, usually during the construction phase.

QUALITY CONTROL GROUP—the group of personnel assigned to implement quality control during the construction phase.

QUALITY CONTROL PLAN—an implementation plan for application of the quality assurance policies during the construction phase.

QUASI-JUDICIAL—a term applied to the action, discretion, etc., of public administrative officers who are required to investigate facts, to draw conclusions from them as a basis for their official action, and to exercise discretion of a judicial nature. The actions of the O.S.H.A. administrators are quasi-judicial in character. When a design professional acts as an arbitrator in resolving disputes between the owner and the contractor, he or she is considered to be acting in a quasi-judicial role. It is in this role that the design professional is granted immunity. (See IMMUNITY)

RECOVERY—obtaining a thing by the judgment of a court as the result of an action brought for that purpose; the amount finally collected or the amount of judgment.

REDRESS—receiving satisfaction for any injury sustained.

REGULATION—a rule or order prescribed for management or government; a regulating principle; a precept; rules of order prescribed by a superior or competent authority relating to the actions of those under its control. An example is the body of federal regulations instituted by O.S.H.A. These regulations must be adhered to by those in the workplace, including the construction project site, or citations will be issued for their violation upon detection.

REJECT ANY AND ALL BIDS—a provision of most invitations to bid for both public and private works. In addition, most jurisdictions grant, by statute or ordinance, that same apparent right to all of its political subdivisions. It is the right of the owner or contracting agency to reject any and all bids, generally for some reason. However, some jurisdictions grant outright authority to reject all bids without cause or for any cause it might deem satisfactory. In some jurisdictions and with some government agencies, it must be shown that the rejection was not arbitrary and capricious. In other jurisdictions, the motive for rejection of all bids is immaterial. Yet, in other jurisdictions, there is the requirement that rejection of bids be predicated on good faith and be exercised promptly. Note that the above deals with the affirmative act of rejection of *all* bids and not with the disqualification of bidders due to material variance in their submission or with the rejection of one bid. In the rejection of a single bid (the lowest responsible and responsive bidder), other factors come into play. In some jurisdictions, the rejected bidder was awarded costs of its bidding process, while in others, though the cost of bid preparation was denied, the contractor was awarded damages to recover reasonable profits, start-up costs, and postbid costs.

RELEASE OF LIEN—the relinquishment, concession, or giving up of the right to a lien by the party in which it exists or to which it accrues. In the construction industry, it is a document releasing the signer's (contractor and/or subcontractor) right to a mechanics' lien on the project.

REPRESENTATIVE—one who stands in the place of another, usually as executor or administrator but not as an agent; one who represents the interests of another. (See AGENT)

RESPONSIBLE BIDDER—one who has the capability, in all respects, to fully perform the contract requirements and the integrity and reliability to assure good-faith performance.

RESPONSIVE BIDDER—one who has submitted a bid under a competitive sealed bid which conformed in all respects to the invitation for bids so that all bidders may stand on equal footing with respect to method and timeliness of submission and as to the substance of any resulting contract. One is responsive if one replies to the specific questions set forth. In the text of public works contracts, one must respond clearly and without qualification to all inquiries addressed to the invitation to bid.

RETAINAGE—an amount of money established by a fixed percentage agreed to in the contract writing that is withheld by one party of the contract from the other as a means of security and/or assurance of performance. In the construction industry, retainage is withheld by the owner against the prime contractor, and the prime contractor, in a similar manner, withholds from its subs. A typical percentage in the construction industry is 10 percent of the amount paid on the progress payments until 50 percent of the work is completed. At that time, the owner may discontinue withholding the retainage.

RISK-SHIFTING TECHNIQUES—typical risk-shifting clauses include indemnification clauses, surety requirements (bid bond, performance bond, and payment bond), "no damage for delay" clauses, etc. Another similar clause is the "condition precedent to payment" clause, which requires the prime contractor to pay his subcontractor only after he has been paid by the owner.

RULES OF DISCOVERY—(See DISCOVERY)

SECURITY—protection; assurance; indemnification; terms usually applied to an obligation, pledge, deposit, etc., given by a party to a contract to the other party. The name sometimes also is given to a party which becomes surety or guarantor for another. In the construction industry, bonds are considered security against default by the bidder or contractor during the respective process.

SHERMAN ANTITRUST LAWS—(See ANTITRUST)

SOLICITATION OF BIDS—(See INVITATION TO BID)

SOVEREIGN IMMUNITY—a concept adopted by the United States from the courts in England, precluding any legal action against public bodies for either breach of contract or for tort claims. This doctrine is applicable at the federal, state, and local levels of government. However, over the last century, this doctrine has waned, especially in the area of tort claims. In most of the 50 states, sovereign immunity is no longer in effect, especially in the area of tort claims. At the federal level, Congress consented to being sued for breach of contract in 1887 by the Tucker Act, and in the tort field, Congress passed the Federal Tort Claims Act in 1946, permitting lawsuits against the United States for certain types of legal wrongs.

STANDARD—general recognition and conformity to established practice; a type, model, or combination of elements accepted as correct or perfect.

STANDARD OF PERFORMANCE—(STANDARD OF CARE)—that standard which a professional (doctor, lawyer, architect, engineer, etc.) must exercise to the degree of care and expertise which a reasonably competent professional of the same discipline would exercise under the circumstances. The standard of performance is established by the professionals working in the same geographical area.

STATEMENT OF PROBABLE CONSTRUCTION COSTS—term introduced by the A.I.A. in 1963 to help minimize the responsibility of guaranteeing the cost estimate. Prior to that time, "cost estimate" was used.

STATUTE—an act of a legislature declaring or prohibiting something; a particular law enacted and established by the will of the legislative department of government. These laws must be adhered to by all parties within that jurisdiction.

STATUTE OF FRAUDS—a statute that requires that no suit or action shall be maintained on certain classes of contracts or engagements unless there shall be a note or memorandum in writing and signed by the party to be charged or by its authorized agent. Its object is to close the door to the numerous frauds and perjuries. In essence, this statute declares that unless a contract is put in writing, it may not be substantiated as legally binding in a court of law. However, one should be aware of the fact that oral agreements are legally binding within certain parameters. These parameters are usually established by the Uniform Commercial Code.

STATUTE OF LIMITATION—(SEE LIMITATION, STATUTE OF)

STRICT LIABILITY—liability without fault. A case is one of strict liability where neither care nor negligence, neither good nor bad faith, and neither knowledge nor ignorance will save the defendant.

SUBCONTRACTOR—a party which takes over portions of a contract from the principal (prime) contractor or another subcontractor; a party which has entered into a contract, express or implied, for the performance of an act with the party which has already contracted for its performance. Most subcontractor contracts hold the subcontractor to the same terms and conditions which are established in the prime contractor's contract with the other parties. Generally, subcontractors specialize in specific building trade, and, as specialists, most subcontractors are licensed by the state in which they operate. The subcontractor's relationship to the prime contractor is that of an independent contractor.

SUBSTANTIAL COMPLETION—the state of completion whereby the building, or a part thereof, is rendered complete to the degree that the owner can use the building, or a part thereof, for its intended purpose.

SUBSTANTIAL CONFORMITY—where a party has complied with the requirements of a writing to the degree that it is essentially the same as that which is required. Substantial conformity might be considered the opposite of material variance. (See MATERIAL VARIANCE)

SUBSTANTIAL PERFORMANCE—exists where there has been no willful departure from the terms of the contract and no omission in essential points; where the contract has been honestly and faithfully performed in its material and substantial particulars, and where the only variance from the strict and literal performance consists of technical or unimportant omissions or defects. In the construction industry, progress payments are made to the contractor based on the substantial performance of the work for that period of time. Usually the issue of substantiality of performance arises when the project is essentially completed, when the owner occupies the building, and when minor deviations from contract requirements become evident. The contractor demands the unpaid balance of the contract price based on substantial performance, and the owner defends by asserting that the balance need not be paid until every deviation is eliminated.

SUPPLEMENTAL CONDITIONS—when an organization has standard general conditions for inclusion in specifications, supplemental conditions are utilized to modify the general conditions to make them project specific.

SURETY—a party which undertakes to pay money in the event that its principal fails. (See BOND)

"SUSPENSION OF WORK" CLAUSE—clause inserted in construction contracts only and which deals with the right of the owner to suspend the work for a period of time as it may determine to be appropriate for the convenience of the owner. When such a clause is inserted into the contract and is then exercised, an adjustment shall be made, an increase in the cost of performance of the contract (excluding profit) necessarily shall be caused by such unreasonable suspension, delay, or interruption, and the contract shall be modified in writing accordingly. However, no adjustment shall be made under this clause for any suspension by the owner if performance would have been suspended by reason of any other cause, including fault or negligence of the contractor, or if an equitable adjustment is provided for or excluded under any other provision of the contract.

SWEAT EQUITY—a term used to mean "mutual help" in certain federal agencies' contracts. The Department of Health and Urban Development (HUD) requires that the tenants of housing built by federal assistance programs, such as housing for the Indians on Indian reservations, contribute to the construction of the unit by giving of their manual labor. This labor is known as mutual help or sweat equity.

TERMINATION—to put an end to; to make to cease; to end.

TERMINATION FOR DEFAULT—construction contracts generally contain specific provisions itemizing events of default. However, even if not specifically itemized, delay in performance resulting in a failure to complete the contract in a timely fashion is universally recognized as a breach of contract. Whether the breach for untimely performance justifies an owner in terminating the contract may depend upon whether "time is of the essence" for performance of the contract. In federal construction contracts, time is of the essence, and if the contractor fails to perform by the date specified, the government may terminate the contract for default. In private contracts, where time is of the essence, the owner has a common-law right to terminate if the contractor fails to perform within the time specified, unless the time for performance has been waived or extended by the acts of the party.

THIRD PARTY—a party which is not privy to a contract but which may be bound or benefited through a written or implied legal relationship.

THIRD-PARTY BENEFICIARY—in order for a party not privy to a contract to maintain an action thereon as a third-party beneficiary, it must appear that the contract was made and intended for its benefit. The benefit must be one that is not merely incidental but must be immediate in such a sense as to indicate the assumption of a duty to make reparation if the benefit is lost.

THIRD-PARTY LIABILITY—a condition whereby a party to a contract may be held liable to a third party related to the contract by its negligent or fraudulent activity in performance of contract. A third party may recover damages where the circumstances are such that the transaction, within the contract requirements, was intended to affect the plaintiff (third party) , and injury to the plaintiff was forseeable.

TORT—a private or civil wrong or injury; a wrong independent of contract.

TORTFEASOR—a wrongdoer; one who commits or is guilty of a tort.

TREBLE DAMAGES—damages given by statute in certain cases, consisting of the single damages found by the jury tripled in amount. The usual practice is for the jury to find the amount of the damages and then for the court to order that amount to be trebled.

TURNKEY CONTRACT—a method of organizing a building project in which a contractor and a designer agree to provide a finished building at an agreed-upon price. Upon completion of the project, all the owner has to do is "turn the key" in the door. Most turnkey projects are built for the Department of Housing and Urban Development (HUD).

UNCERTAINTY—a state or quality of being unknown or vague; such vagueness, obscurity, or confusion in any written instrument, *e.g.*, a contract, as to render it unintelligible to those who are called upon to execute or interpret it so that no definite meaning can be extracted from it.

UNCONSTITUTIONAL—that which is contrary to the constitution. The term can be used in two different senses. The first is that legislation conflicts with some recognized general principle or conflicts with a generally accepted policy. The second is that the legislation conflicts with some provision of the written constitution which it is beyond the power of the legislature to change.

UNIFIED BID—in a multiple prime construction contract, solicitation for bids is presented to the bidders in several separate prime contract packages. In a unified bid procedure, the bidders are permitted to bid on either one or as many of the bid packages as are presented.

UNIFORM COMMERCIAL CODE—a body of laws which governs the sale of goods in almost every state of the United States. Application of the Uniform Commercial Code in the construction industry is rare except in the area of shipping, handling, and purchasing of materials for the project.

UNJUST ENRICHMENT—doctrine stating that persons shall not be allowed to profit or to enrich themselves inequitably at another's expense. A typical example of this doctrine is when an owner withholds payment to a contractor for work already performed, claiming that work is not acceptable. The value of the performed work far exceeds that portion which the owner considers unacceptable.

VARIANCE, MATERIAL—(See MATERIAL VARIANCE)

WAIVER—the intentional or voluntary relinquishment of a known right. Waiver is essentially unilateral, resulting as legal consequence from some act or conduct of parties against which it operates, and no act of the party in whose favor it is made is necessary to complete it. In the construction industry, an owner may waiver his or her right to a signed change order for work incorporated into the project when the following conditions exist: the owner is aware of the change but does not object; the item is of such magnitude that the change could not be made without the owner's knowledge; the changes are necessary but were not foreseen by the design professional; and some subsequent oral agreement (the architect's orally approved substitution and/or change with which the owner agreed) waives the requirement of a signature.

WARRANTY—a promise that a proposition of fact is true.

WARRANTY OF SPECIFICATIONS—the owner's implied warranty, when providing the plans and specifications, that the plans and specifications are possible to perform, are adequate for their intended purpose, and are free from defect. Moreover, this warranty is not overcome by the usual exculpatory clauses requiring bidders to visit the site, check the plans, or generally inform themselves of the requirements of the work. The implied warranty of specification suitability has been recognized in every American jurisdiction and applied with equal force to public and private contracts. As with any contractual obligation, the warranty can be overcome by explicit contractual provisions which impose absolute liability for performance on the contract.

ZONING—a local ordinance which governs the uses of land and the overall characteristics of the structures that may be erected; the division of a city by legislative regulation into districts and the prescription and application in each district of regulations having to do with structural and architectural designs of buildings and of regulations prescribing uses for buildings within designated districts.

Additional Construction Documentation Resources

AMERICAN JURISPRUDENCE

13 Am. Jur. 2d *Building and Construction Contracts*

13 Am. Jur. 2d *Buildings* , §§ 8–11

17 Am. Jur. 2d *Contractors' Bonds,* §§ 1–4

41 Am. Jur. 2d *Indemnity,* § 9 *et seq.*

41 Am. Jur. 2d *Independent Contractors*

57A Am. Jur. 2d *Negligence,* § 51

66 Am. Jur. 2d *Restitution and Implied Contracts,* §§ 21, 67–68, 70–71

CORPUS JURIS SECUNDUM

17A C.J.S. *Contracts,* §§ 498(2)–532

42 C.J.S. *Indemnity,* § 5

AMERICAN LAW REPORTS

12 A.L.R. 1409 (1920)—What is "accident" within provision of bond or contract indemnifying against damage or injury to person or property by accident in performance of building or construction contract.

38 A.L.R. 566 (1923)—Liability of contractor and contractee inter se in respect of damages paid to third persons for injuries sustained during the progress of the stipulated work.

40 A.L.R. 928 (1925)—Revocability of municipal building permit or license.

53 A.L.R. 103 (1927)—Who must bear loss from destruction of or damage to building during performance of building contract, without fault of either party.

64 A.L.R. 920 (1929)—Validity and construction of zoning or building ordinance prohibiting or regulating subsequent

alteration, addition, extension, or substitution or existing buildings.

107 A.L.R. 960 (1935)—Construction and application of provision of construction contract as regards retention of percentage of current earnings until completion.

107 A.L.R. 1035 (1935)—Owner's right to rescind building and construction contract for default of contractor.

121 A.L.R. 345 (1939)—Legality of combination among building or construction contractors.

129 A.L.R. 534 (1939)—Statute of frauds against oral contracts not to be performed within year as applicable to contracts susceptible by its terms, or by construction, of performance within year, where performance within that time is improbable or almost impossible.

6 A.L.R. 2d 960 (1948)—Rights of permittee under illegally issued building permit.

22 A.L.R. 2d 647 (1950)—Right of owner's employee, injured by subcontractor, to recover against general contractor for breach of contract between latter and owner requiring contractor and subcontractors to carry insurance.

59 A.L.R. 2d 885 (1957)—Validity of statute making private property owner liable to contractor's laborers, materialmen, or subcontractors where owner fails to exact bond or employ other means of securing their payment.

82 A.L.R. 2d 1429 (1959)—Failure of artisan or construction contractor to procure occupational or business license or permit as affecting validity or enforceability of contract.

2 A.L.R. 3d 620 (1960)—Effect of stipulation, in private building or construction contract, that alterations or extras must be ordered in writing.

6 A.L.R. 3d 1394 (1963)—Construction contractor's liability to contractee for defects or insufficiency of work attributable to the latter's plans and specifications.

25 A.L.R. 3d 383 (1968)—Liability of builder-vendor or other vendor of new dwelling for loss, injury, or damage occasioned by defective condition thereof.

26 A.L.R. 3d 1395 (1968)—Failure of artisan or construction contractor to comply with statute or regulation requiring a work permit or submission of plans as affecting his right to recover compensation from contractee.

50 A.L.R. 3d 596 (1972)—Retroactive effect of zoning regulation, in absence of saving clause, on pending application for building permit.

61 A.L.R. 3d 792 (1970)—Liability of builder or subcontractor for insufficiency of building resulting from latent defects in materials used.

61 A.L.R. 3d 1128 (1973)—What interest qualifies one as an "owner" for purposes of making application for a building permit.

65 A.L.R. 3d 249 (1973)—Tort liability of project architect for economic damages suffered by contractor.

10 A.L.R. 4th 385 (1980)—Liability of builder of residence for latent defects therein as running to subsequent purchasers from original vendee.

12 A.L.R. 4th 866 (1978)—Statutes of limitation: actions by purchasers or contractees against vendors or contractors involving defects in houses or other buildings caused by soil instability.

American Jurisprudence Legal Forms

Building and construction contracts, and specifications of requirements and work methods—2 Am. Jur. Legal Forms 2:1063-2:1138

Contract provision requiring contractor to carry liability insurance—2 Am. Jur. Legal Forms 2:1242, 2:1243

Contractual provisions as to assignment and subletting—2 Am. Jur. Legal Forms 2:1118-2:1124

Contractual stipulations as to governing law—2 Am. Jur. Legal Forms 2:1147

Contractual stipulations as to right of owner to terminate or cancel contract—2 Am. Jur. Legal Forms 2:1127, 2:1240, 2:1241

Indemnity against negligence of indemnitee—6 Am. Jur. Legal Forms 6:30

Provisions for reimbursement for wrongful act or neglect of other party—2 Am. Jur. Legal Forms 2:1128

Provisions permitting alterations and modifications—2 Am. Jur. Legal Forms 2:1152-2:1155

Provision that all alterations or extras shall be in writing—2 Am. Jur. Legal Forms 2:1191-2:1193

Stipulation as to independent contractor relationship—2 Am. Jur. Legal Forms 2:1148

Stipulation that terms of contract to control in case of conflicts with proposal—2 Am. Jur. Legal Forms 2:1145

Acceptance of proposal by general contractor—4 Am. Jur. Legal Forms 2d 47:38

Acceptance of substantial performance—4 Am. Jur. Legal Forms 2d 47:301

Architect's or engineer's certification effect of certificate in regard to architect's or engineer's liability—4 Am. Jur. Legal Forms 2d 47:307

Bond provision—Purpose—To protect owner—5A Am. Jur. Legal Forms 2d (rev. ed.), Contractor's Bonds 67:91

Building permit as condition to sale—Application—16 Am. Jur. Legal Forms 2d 219:540

Contract for construction of building—4 Am. Jur. Legal Forms 2d 47:56 *et seq.*

Contractor's performance bond—5A Am. Jur. Legal Forms 2d (rev. ed.), *Contractor's Bonds* 67:81, 67:83

Labor and material payment bond—5A Am. Jur. Legal Forms 2d (rev. ed.), *Contractor's Bonds* 67:131-67:133

Owner's protective bond—5A Am. Jur. Legal Forms 2d (rev. ed.), *Contractor's Bonds* 67:81

Provision for notice—13A Am. Jur. Legal Forms 2d 183:12

Contract to demolish building—2 Am. Jur. Legal Forms 2d 49:16

Nonperformance of contract—9 Am. Jur. Legal Forms 2d 142:34

AMERICAN JURISPRUDENCE PLEADINGS AND PRACTICE FORMS

Answer alleging work not done in workmanlike manner—5 Am. Jur. Pl. & Pr. Forms 5:562

Complaint by contractor to recover for additional work—5 Am. Jur. Pl. & Pr. Forms 5:555

Index

A

B

Q

R

S

T

U

V

W

Z